"The spirit of Greta Thunberg pulses through these pages. More than a how-to guide, *Tiny Transit* is a joy to read."

– Linda Wurzbach
President, Resources for Learning
Austin

"Amory Lovins famously said, 'Energy conservation isn't a free lunch. It's a lunch *we pay you* to eat!' And now Susan Engelking has served up a banquet in her inspiring new book, *Tiny Transit*.

Engelking shows how innovation is already blooming coast-to-coast as cities open the way for small, low emission vehicles to expand mobility, reduce congestion, and advance safety – all at a fraction of the cost of conventional infrastructure."

– Professor Michael Wheeler
Harvard Business School

"This is a real page turner. I have never seen the incredibly complex subject of microtransit laid out there so succinctly."

– Chris Nielsen
Founder and CEO
Electric Cab North America

"Just imagine if all of the public parking for cars were turned into *actual* parks, or parking and lanes for cleaner, greener mobility options. *Tiny Transit* encourages readers to see their cities through a different lens."

– Michael Keating
Founder & President, Scoot

"Susan Engelking takes readers' hands and walks them to a better future."

– Lisa Kay Phannenstiel
Executive Director, Movability
Austin

"With twenty more initiatives of this scale, we can tip the course of history."

– Joan Marshall
Director, The Bryan Museum
Galveston

"As young Greta Thunberg urges us to action on the climate crisis in general, Susan Engelking's *Tiny Transit* strategy is one concrete solution to the burgeoning crisis of car, bus and truck-choked cities."

– Pamela Ryan, PhD
Author, *Impact Imperative:*
Innovation, Entrepreneurship and
Investing to Transform the Future

"Imagine what is possible when children are encouraged to become climate innovators."

– Cristal Glangchai, PhD
Founder, VentureLab
Author, *Venture Girls: Raising Girls*
to be Tomorrow's Leaders

Tiny Transit™

Tiny Transit™

Cut Carbon Emissions in Your City
Before It's Too Late

Susan Engelking

NEW YORK

LONDON • NASHVILLE • MELBOURNE • VANCOUVER

Tiny Transit™

Cut Carbon Emissions in Your City Before It's Too Late

© 2020 Susan Engelking

Published in New York, New York, by Morgan James Publishing in partnership with Difference Press. Morgan James is a trademark of Morgan James, LLC. www.MorganJamesPublishing.com

ISBN 9781642796827 paperback
ISBN 9781642796834 eBook
ISBN 9781642796841 audiobook
Library of Congress Control Number: 2019943707

Cover and Interior Design by:
Chris Treccani
www.3dogcreative.net

Morgan James is a proud partner of Habitat for Humanity Peninsula and Greater Williamsburg. Partners in building since 2006.

Get involved today! Visit
MorganJamesPublishing.com/giving-back

For Tyler, Jack, Courtney, Dane, and Nan

Look at the world around you. It may seem like an immovable, implacable place. It is not. With the slightest push — in just the right place — it can be tipped.

– Malcolm Gladwell

The Tipping Point:
How Little Things Can Make a Big Difference

Table of Contents

Foreword

Young people today want to live near urban centers, to be part of vibrant communities, to be lifelong learners, and to experience life. They are a sharing generation that looks for new, less expensive, sustainable lifestyle options – from cars to homes.

You bet young adults are deeply concerned about climate change. They are conscious of their carbon footprints, and they truly want to leave this planet better than they found it.

Imagine the world this fresh generation foresees, where people can get everywhere they need to go within our communities using environmentally friendly, low-cost, safe options.

This is not a dream. Innovative cities from California to Georgia are creating networks with off-road paths for low speed electric vehicles without huge infrastructure costs. There is an emerging groundswell of people who see the practicality and advantages of this approach. With low-speed networks, cities can become truly multimodal so that people can choose and blend their best options based on their budgets and where they need to go.

Tiny Transit offers an effective and logical way to respond to what young people want. Low-speed networks create possibilities for people – to get better jobs, to go to classes, to go places easily and safely, and to live without the anxiety of extreme financial stress. Susan brings this movement directly to you. She walks you step-by-step through how to introduce this new mobility alternative, help people envision change, create the infrastructure for less than you might imagine, and even melt opposition. With low speed networks, cities can become truly multimodal so that people can choose and blend their best options based on their budgets and where they need to go.

What I particularly like about this book is that air quality is merely the starting point for learning so much more about the benefits of Tiny Transit™. I hope city leaders everywhere will join in this exciting new approach to creating more livable communities.

— Nan McRaven, EdD

Chapter 1:

Introduction

This book is for cities that want to cut carbon emissions.

It is not a book about climate change. There are excellent books and reports on that subject.

Tiny Transit is a how-to guide for cities, Councils of Governments (COGs), Metropolitan Planning Organizations (MPOs), and other regional planning organizations that are searching for ways to help their member cities improve mobility while reducing carbon emissions quickly and inexpensively.

COGs and MPOs are how cities, towns, and counties are organized to coordinate in their collective best interests on transportation, air quality, economic development, and more. You are their resource on matters of mobility, air quality, and carbon emissions, especially for smaller cities that don't have this expertise within their staff. They look to you for expertise and guidance. That's why I am speaking directly to you.

Your role is vital. There are 35,000 cities and towns in the U.S. They are too numerous for me to reach them all directly with this how-to guide. You are the link, the key to introducing to your cities the low emission mobility alternative set forth in this book.

The Environmental Protection Agency (EPA) and U.S. Department of Transportation see this tiered relationship with your cities similarly. That's why these agencies direct resources to organizations like yours to improve air quality and mobility.

In this book, you will learn about a proven, clean, low-cost, fun mobility alternative that cities across the U.S. can quickly implement to dramatically reduce carbon emissions – a public health hazard and the number one cause of climate change – and build a groundswell of public awareness and support in the process.

LEAN Networks and Low Emission Modes

Let me explain the key terms you will encounter in this book.

Throughout, you will learn about LEAN networks. The term LEAN is derived from *Low Emission Alternative Network*. LEAN networks are the infrastructure that provides safe passage for a variety of low speed, low emission, low cost modes. LEAN networks are specifically designed for Neighborhood Electric Vehicles (NEVs), scooters, pedicabs, bicycles and pedal assist bicycles, and other low speed modes.

Here are some other terms you will encounter:

- *LEAN Lanes*™ are thoughtfully designed as protected routes for these low speed modes. A LEAN Lane could be an off-road path or a distinct, separated route on an arterial road. Connect these lanes, and cities create protected networks that accommodate low speed modes.
- *Protected networks* separate low speed modes from conventional vehicles for reasons of public health and safety.

- *Low speed modes* are designed not to exceed twenty-five mph no matter how hard you press the pedal.
- *Neighborhood Electric Vehicles (NEVs)* are not golf carts despite their appearance. Rather, NEVs are a distinct classification of vehicles, also called Low Speed Vehicles (LSVs), that are required by law to meet stringent federal standards that make them street legal in most states, unlike golf carts. NEV is usually pronounced *en-ee-vee*, but some Californians pronounce it *nevs*.
- *Tiny Transit*ᴛᴍ refers collectively to these low speed, low emission modes.
- *Dockless* typically refers to shared bicycles and scooters that can be unlocked, ridden, and then locked again using a phone app without a docking station.
- *Redundant safety* means not relying on one road feature alone to protect low speed modes. With redundant safety, several things need to go wrong for someone to be injured.
- *Micromobility* is becoming a popular term for the various low speed, low emission approaches – both the infrastructure and the modes that use the infrastructure, like NEVs.
- *Innovator and early adopter cities* are lighting the way for us all. In this book, you will learn from cities that already incorporate NEVs in their strategic mobility plans, including cities that have more than a decade of experience with NEVs.

The Cost of Nonattainment

Approximately fifty areas in twenty-two states are in nonattainment, meaning that their air is unsafe for many people to breathe. The effects are life-threatening, even deadly.

Miners actually used to carry caged canaries while working deep underground. The canaries were the way that miners measured air quality. If the canary died, it meant that there were dangerous gases and that the miners should get out as quickly as they could.

Nonattainment designation is truly today's "canary in a coal mine." It warns cities that air pollution is at unhealthy, even dangerous levels. The miners could scramble out, but the only solution for U.S. cities is to correct their course.

Many attainment cities are hovering on the brink of nonattainment. In July 2018, San Antonio finally lost its battle to meet federal ozone standards.

> Levels of smog-producing ozone in San Antonio's air have finally triggered what local offices have long feared – a designation of nonattainment by the Environmental Protection Agency, which could hurt the city's economy by delaying transportation and manufacturing projects. Now any future highway expansions – which produce more pollution from traffic – will likely be delayed for years. 'For fifteen years, we fought it and survived, but it looks like our day has come,' said Mayor Ron Nirenberg. 'The city is planning a mass transit system to cut back on greenhouse gas emissions by reducing traffic.' … [Former mayor Nelson Wolff] said, 'Public transit's going to be the key.'
>
> **– John Tedesco,**
> "Ozone levels trigger EPA nonattainment designation in San Antonio," *My SA*, July 18, 2018

Calling All Innovators

The critical path is fast action by cities. We don't have the luxury of time. We don't have the luxury of process. When it comes to carbon emissions, we cannot shoot for 2030. We simply can't wait another decade or two to wean ourselves from fossil fuel. We can't wait five years. It has to be now.

With this book, I am casting a line in the water for prospective innovators and early adopters, cities who will join the cities that are already trying, testing, piloting, and implementing protected networks for low speed, low emission mobility modes.

As you read through, I encourage you to think, which of your member cities are prime candidates to join these innovators and early adopter cities? Your prime candidates could be cities that are on the brink of nonattainment, whose leaders see that they must change course without delay, or your candidates could be any cities that simply want cleaner air or that want to be part of the solution to climate change.

Good news – this isn't one of those take-your-bitter-medicine approaches. It's entirely voluntary, it can save people a lot of money, and it can add joy to people's lives.

No one has to give up his car. No one has to do anything, really. But people will want to, I promise. I'm seeing it. This alternative is so practical, inexpensive, and fun that it's a people-magnet. You don't have to sell people on this idea. All you have to do is show them.

Steve Jobs said, "Our job is to figure out what customers are going to want before they do...." In that spirit, how can people want to drive a Neighborhood Electric Vehicle if they have no idea that it even exists?

This approach takes resources, yes. But they are a fraction of what you might imagine. I will introduce to you a paradigm change that turns what we think we know about transportation on its head.

Wouldn't it be great if people were actually clamoring to reduce their carbon emissions? Can you envision a world where people are smiling, ready, and eager to become part of the solution to climate change?

Imagine that in your region and all across America, cities and communities are working to make mobility safer, kinder, gentler, cheaper, more fun, and free of carbon emissions. As innovator cities develop successful, popular demonstration projects, other cities will be confident in joining them.

If you are working with cities that want carbon emission solutions with impact, solutions that can be quickly implemented, *Tiny Transit* is for you.

Benefits Jackpot

Until this point, I have focused on cutting carbon emissions. Now I want to prepare you for an even broader perspective. You will soon learn that LEAN networks can benefit cities in a multitude of ways, including:

- Public safety
- Public health
- Economic development
- Economic opportunities
- Social equity
- Wheelchair user access
- Affordability
- Fiscal responsibility
- Economic resilience

With the LEAN approach, you have hit the benefits jackpot.

Quick Start Guide

In the coming pages, you will learn more about these benefits and how they can attract allies. I call the collection of diverse allies your "big tent." They are the key to accelerating this new mobility alternative – and getting it right.

The final chapter is your Quick Start Guide. You'll be presented with a game plan that you can adapt for your unique city, region, organization, or member governments.

I'll also make myself available to you. I'm interested in what cities are doing across the country, and I look forward to connecting with you.

Chapter 2:

My Origins

Before we start, I would like to share with you how I arrived at this point.

Even in elementary school, I was thinking about our country's dependence on gas-guzzling cars.

In high school, I believed we wouldn't be driving internal combustion engines by the time I was an adult. An Arab oil embargo made clear the high price of our dependence on oil. It just seemed insane to use one gallon of gasoline – a precious resource that was 300 million years old – to go to the grocery store and run errands.

I thought everyone could see this; it was so obvious to me. But I came to understand that there weren't many people who saw it. Even my friends at school had puzzled looks on their faces when I would make some reference to "when we don't drive cars anymore."

I thought about transportation a lot more than the average person. My father was a civil engineer who designed highways. We went on Sunday drives on stretches of highways that weren't yet open. He would edge past barricades to drive our family around and around a new cloverleaf until we were dizzy. He would terrify us all by speeding up to a portion of a bridge that wasn't finished and stop suddenly at the edge, just short of sending us hurtling off into an abyss. It terrified my mother, but my sisters and I loved it.

All that is to say that I thought of roads and cars not just *more* than the average person. I thought of them *differently*. I had no interest or aptitude for engineering, but I thought of roads as things that people actually designed and built, and I believed that *they could be different.*

At the age of fourteen, I had my mother drive me to my first protest march. There was an enormous city-wide conflict in San Antonio over a proposed expressway to connect downtown with an outer loop, a mile from where we lived. The problem was that there was not enough room for a freeway. It would divide neighborhoods, condemn and demolish stately old homes, and – worst of all in my view – consume the land next to my beloved San Antonio Zoo. It would have cars spewing exhaust run right next to the animals – elephants, monkeys, lions, and giraffes. As if it wasn't enough that they had to live in small cages, they would now have to breathe noxious fumes that would make them ill and shorten their lives.

The McAllister freeway was built, of course.

At fifteen, I placed second in a regional extemporaneous writing competition. My subject was the crisis of air pollution.

Fast forward to age twenty-two and the first job where I wasn't scooping ice cream. I became aide to an Austin City Council Member who later became mayor. His focus was on economic development, historic preservation, and revitalizing downtown – already interests of mine. I fell in love with Jane Jacobs and her groundbreaking book, *The Death and Life of Great American Cities*. I learned that she had fought against and defeated an expressway proposed by Robert Moses that would have cut through Manhattan's Greenwich Village. I marveled at how she built a grassroots movement that changed how we think about city life. I saw the connections between what happened at City Hall and our built and natural environments. I became aware of the heavy costs of sprawl. I had my first brush with wily special interests.

Most of all, I came to believe in cities and neighborhoods as an instrument of change.

Stirrings of Discontent

In 2005, when my son Tyler was a year old, I realized that the biggest threat to his life and health wasn't getting into the knife drawer or toppling into the toilet. It was riding in a car in Austin, Texas.

During college, I lost two friends I'd grown up with. Debbie was killed when the car she was driving spun off a highway in Alaska. Jim was killed in a head-on collision on a narrow mountain road in Colorado. Both his mother and sister had been killed in a crash a few years earlier.

I knew five people whose *children* were killed in car crashes. Two of these friends died young themselves, I believe from grief.

In 2005, I began researching the problem. I found that more than seventy people in Austin are killed each year in car crashes. More than 3,700 are killed on Texas roads every year. One-third of those are bicyclists, motorcyclists, or pedestrians. More than 37,000 people are killed on our nation's roads every year. The numbers fluctuate year to year, but they are fairly constant.

I learned that Houston, Texas area roads are the most dangerous in the U.S. according to the *Houston Chronicle*. The number of people killed on its roads are the equivalent of three fully loaded 737s crashing each year, killing everyone aboard.

Yet there is no equivalent to the FAA or CDC to swoop in and analyze what happened. There is only silence and sadness. Where is the outrage?

For every fatality, there are four people with life-altering injuries – brain injuries, paralysis, loss of limbs, chronic pain, injuries from which they may later die. In Texas, more than 17,000 people suffer serious injuries each year.

Families devastated, lives shattered, lost wage earners, bankruptcies, poverty, painkiller addiction – the cascade of miserable consequences never stops. All for what?

Why do we accept this level of carnage?

I decided, *this has to change.* I resolved that if the transportation industry could not solve the crisis of roadway fatalities by the time that Tyler was old enough to drive, I would solve it myself.

Of course, I really thought that experts would solve the problem. I was giving them fifteen years. Who was I even to think of solving it?

In the meantime, I started learning more about traffic fatalities. For the next nine years, I studied the issue, creating matrices for each year listing every person who was killed on Austin roads – age, ethnicity, crash conditions, where and how they died.

I interviewed experts. It's amazing who will meet with you if you just call them on the phone and ask.

One of the most disturbing interviews was with the head pediatric surgeon at a local hospital emergency room. His dark eyes were piercing as he told me about the children and teenagers who arrived broken or dead from car crashes. He believed that every one of those injuries and deaths was preventable. I felt his fury as he told me that he had repeatedly warned highway officials about a dangerous road where a child was killed but was told that the money to fix it wasn't in that year's budget or the next's. No one acted on his warnings, and within months, another child was killed at the same place.

Helena, a high school girl who aspired to someday be the headmistress at a Montessori school, was killed on a dangerous curve one drizzling Saturday morning while driving to her Advanced Placement tests. Her heartbroken mother wrote to me about the dangerous curve where her daughter was killed. "There have been efforts to get it straightened out, but no one has succeeded yet. It's very dangerous. My husband hydroplaned on that exact same spot when Helena was an infant, rolling the van he was in with both my daughters in the vehicle. No one was injured. After Helena's death, we learned that three people have died on that curve in the past four years."

The more I learned, the stronger my resolve became. By now, I had another reason to become a safety advocate: Tyler's little brother Jack was born.

At this point, a staff member at the Austin Police Department called, wondering why I had requested so much data on traffic fatalities. To my surprise, the police wanted to work with me. They hoped I might be the change agent this public safety crisis desperately needed.

I met with Police Chief Art Acevedo, then in Austin, now chief of police in Houston. Even as we spoke, he called staff to come in and listen to our discussion. He liked my recommendations and encouraged his staff to work with me.

I learned that the police show up at every injury crash, while the traffic engineers who design roads are in their offices, working on their CAD programs, or sound asleep at home. It's the police who deal with the consequences of their designs – scraping bodies off the pavement, interviewing witnesses, notifying next of kin. There is a dysfunctional distance between the two.

I became increasingly knowledgeable about roadway safety. My friends were among the first to learn about Vision Zero, the movement that began in Sweden to end roadway fatalities, because I told everyone that's what we needed to copy. I probably spent too much time talking about gory, gruesome crash details with friends. Who wants to hear about that over dinner? No one!

I began going to forums, even speaking to a couple of groups about my findings and recommendations. I got to know wonderful activists and organizations like Mothers Against Drunk Driving, better known as MADD.

So I was busy doing things.

But I wasn't making a difference. Nothing was changing. Auto makers couldn't come up with safety features fast enough to keep up with increasingly distracted drivers. I learned that even a car that's built like an armored tank can't save someone who is hurtling into a crash going sixty-five mph on impact. All the airbags in the world can't save internal organs from damage in a high-speed crash.

Again, why do we tolerate this carnage on our roads? Lethargy. Ignorance. Our society has a high tolerance for roadway deaths and injuries. We are desensitized.

"It was an accident," people say sadly, as if a death couldn't have been avoided.

In time, the idea of Vision Zero spread to the U.S. But as I met with some transportation officials, it was clear that they could talk about it, but they weren't actually committed to it because they didn't believe it was possible. Vision Zero was seen as a vague, unachievable, distant, worthy aspiration, not an urgent mandate for change.

On Tyler's tenth birthday, I realized that he was in just as much danger on our roads as when I started.

Remember the story of Sleeping Beauty? When she was an infant, her mother, the Queen, invited the fairies to her christening. But an uninvited evil witch-fairy showed up and cursed the infant: "When you are sixteen, you will injure yourself with a spinning wheel and die!" The girl grew up as a lovely princess. Her parents banished all spinning wheels from the castle. They did everything possible to get rid of every spinning wheel in the kingdom. But on her sixteenth birthday, the curious girl encounters a spinning wheel, pricks her finger, and falls into what today we would call a coma, in spite of her parents' best efforts.

I thought of my quest's parallel to Sleeping Beauty. I was determined to protect Tyler and Jack from car crashes, and my deadline was the same as Sleeping Beauty's – Tyler's sixteenth birthday, the day he would be old enough to drive. No matter how much I tried to protect him, I knew that danger was lurking out there as surely as spinning wheels were lurking in the castle's hidden rooms.

Every time my energy waned, I reminded myself that if I could just crack the code on roadway fatalities, it would save thousands of lives. No one would ever know whose lives were saved, but one of them might be Tyler's or Jack's. That commitment, an inner fury, kept me going.

A Collaboration Is Born

During the years before Tyler and Jack were born, I was project manager for a long-range economic plan for Austin that identified information technology as a growth sector while the Internet was in its infancy. Where Austin's primary emphasis had been on recruiting corporations, it shifted to improving Austin's quality of life, education, and economic opportunities to become a magnet for entrepreneurial talent. I worked with SRI International and the team of community leaders who developed this holistic, winning approach.

For Austin's next two long-range economic plans spanning two decades, I served as senior editor.

It was while I was working on an economic development project for Baltimore City that I first thought of a hazy concept I called Tiny Transit. I instantly saw it was at least a bronze bullet for on-demand mobility, carbon emissions, social equity, economic resilience, economic opportunity, fiscal restraint, affordability, tax base, public safety, public health, and more.

Baltimore's air quality was in nonattainment. I saw the implications for clean air and climate change. I was blown away.

I asked my boss to stop by my office and sketched the idea. He closed the door and told me not to tell anyone about it. I was confused. Weren't we paid to come up with breakthrough ideas?

"This idea is so good it will hijack our company," he said. "It's all anyone will want to work on." At that moment, I knew I had hit upon something.

Baltimore passed on the idea, but I was in its thrall. I talked with everyone about it – friends, acquaintances, and people I met in line at Whole Foods. Several suggested that I approach the City of Austin about it. I did, briefly and unsuccessfully. I wasn't surprised. I'd worked for the City Council. I feared it would be a frustrating, years-long battle – one I had no interest in fighting.

Then a long-time friend suggested that I call Professor Michael Walton at the University of Texas at Austin, a renowned engineer and transportation consultant. I did, and he invited me over the next day. I could hardly believe how accessible he was. I handed him a draft of the idea and watched as he read it and began to smile.

"You have to meet Katie Kam," he said. He had supervised Katie's dissertation in civil engineering, a feasibility study on a concept similar to what I was describing. Her data proved that low speed vehicles continuously flowing at twenty-five mph, unimpeded by traffic signals, could get places as fast as or faster than conventional vehicles at peak hours in Austin traffic.

Katie and I met and immediately rolled up our sleeves. We've never rolled them back down.

Katie not only has a Ph. D., she also did her post-doctoral work at the University of Texas Center for Transportation Research. She began her career as a city planner and earned two master's degrees, one in community and regional planning from the UT School of Architecture, the other in civil engineering. She is a board-certified Professional Engineer who casually uses terms like "ingress" and "egress."

Katie had studied what other cities were doing to incorporate Neighborhood Electric Vehicles in their plans. She had reviewed their plans and talked to early adopters as part of her research. So we began our collaboration with a proven concept.

The key to "getting there faster by going slower" is continuous flow, achieved by incorporating bridges, tunnels, and smart design so that the NEV drivers are flowing unimpeded at twenty-five mph. At that speed, they are passing drivers of conventional vehicles on slow moving, congested roads.

Katie is a thoughtful transportation advocate. Her civil engineering firm is Wheels & Water LLC, with a particular interest in developing and implementing safer, more efficient transportation options, especially for NEVs. It was Katie who hit upon the idea of "lean mobility."

For two years, we met nearly every Sunday to work on the concept and then set up appointments during the work week. No one paid us to do this. We both made sacrifices to pursue our passion, so certain were we of its potential.

Plus, we both had children, and we were acutely aware of how dangerous our roads are for inexperienced young drivers.

In a matter of two years, Katie and I met with more than one hundred people. I stopped counting. I can honestly say that for at least the first fifty meetings, I learned things in every meeting that affected my thinking and our concept.

In 2017, our application to raise awareness of the LEAN concept was one of 70 out of 300 selected for an Innovation Showcase at a Smart Cities Connect conference by US Ignite, organized by the Office of Technology and Policy, led by the National Science Foundation. Our table was so popular that a man at a nearby transportation table tried to hire me to run his.

That event opened doors. Among the people I met was the director of transportation for the City of Raleigh, who is now director of transportation for the City of Dallas. I anticipate that Dallas will include LEAN networks in its forthcoming strategic mobility plan.

In recent months, both Katie and I have intensified our efforts. Our mission is now to *help cities develop a groundswell of support* for protected networks for low speed, low cost, *low emission mobility alternative*. Cities, we believe, are the point source for change.

Tyler Is Struck

On April 20, 2018, my son Tyler was struck by a car while he was on his bicycle in a crosswalk. It was a four-lane road with a grassy median. He'd gotten to the median and was waiting for the remaining two lanes of traffic to stop. The car in the lane closest to him stopped and waved him through. Tyler started across, but the driver in the next lane didn't see him until too late. The first car sped off and was never identified. Tyler was knocked unconscious. Bystanders pulled him to the side of the road and called 911. One man took off his shirt to wrap Tyler's bloody body. He was rushed to the Dell Children's emergency room.

Tyler lived. He was released after five hours. He was bruised and bloodied, he limped for months, he went to physical therapy, and he missed a week of school, but he lived, his body intact. His leg still hurts at times, but he doesn't complain.

Seared in my mind is an image of his slender ninety-five-pound body struck by a mass of steel weighing 4,000 pounds.

I have shared my story with you so that you understand the LEAN approach. The LEAN concept is not a whim. It has been gestating for over a decade, developed in collaboration with a PhD civil engineer and urban planner who has studied this alternative, and vetted with serious experts. It is based on knowledge, not naiveté.

Here's how it all comes together to help you: Your mandate is to improve mobility while reducing carbon emissions. It sometimes feels like swimming against the current. I understand the urgency and difficulties you face.

With all of that, how do you get people to change?

By offering people a mobility choice that is not only cleaner for our air but also safe, inexpensive, and effective in getting people to their destinations fast. So let's get started.

Chapter 3:

Leapfrog to a Low Carbon Emission Future

What if we throw out everything we think we know about transportation? What if we set aside our preconceptions about transit? What if instead of thinking *big*, we think *small?* Think for a moment about historic shifts that swiftly swept entire industries.

- Thirty-five years ago, half of the world had never used a phone. Everyone knew that we couldn't extend telephone lines to the entire planet. Then came wireless technology, which leapfrogged landlines. Today you can probably find a goat herder in Outer Mongolia with a cell phone the size of a deck of cards.

- Forty years ago, computers were huge, expensive, and centralized. Then came Steve Jobs, Steve Wozniak, and Apple to radicalize computing and give power to the people, creating industries where none existed and products that are now woven into our everyday lives.

Transportation is overdue for a transformation. Our inability to break out of 1950s-era urban design thinking and wean ourselves from the internal combustion engine and conventional-size vehicles not only threatens our communities' health and safety, our national security, and our economic resilience, it also now threatens life on our planet.

Transition Time

What if cities could leapfrog our way into the future of transportation? What if we approached mobility in the way that Jane Jacobs approached urban planning? What if it isn't about fixed routes and limited choices? What if we redefine transit, with people going where and when they want to go? What if mass transit centers were connected with micromobility modes? Envision a world in which people are eager to become part of the solution.

What if these small, vulnerable modes weren't sandwiched between multi-ton SUVs and trucks, praying to be seen? What if instead you could step out of a building and into a Neighborhood Electric Vehicle (NEV) on your own schedule? What if you could zip along on narrow lanes dedicated to small vehicles with maximum speeds of twenty-five mph? What if you could hop in a tiny vehicle and zip to a light rail station? What if parking was convenient and cheap, or better yet, free?

The transition can happen quickly. If you'd told people at the turn of the twentieth century that the horse and buggy would soon be history, they might have laughed. A few years later, they were behind the wheel.

Let me reassure you. We are not talking about a wholesale shift away from conventional cars and trucks. They will continue to constitute the vast majority of trips – say sixty to eighty percent – for some years to come. Katie's dissertation, "The Transition to Low Speed Vehicles for Intra-City Travel," underscores that the next ten to twenty years will be a time of transition – not of revolution but rapid evolution.

The sooner cities undertake the transition, the faster it can happen and the faster cities can cut carbon emissions. What are we waiting for?

LEAN Networks, A Gentle Disruptor

LEAN networks are at the intersection of *carbon emissions* and *traffic fatalities and life-altering injuries*. By ensuring safety, LEAN networks enable low emission modes to gain traction among the public. People will choose these modes – for themselves, for their teenage drivers – only if they are safe. LEAN networks provide that safety.

A LEAN network is a protected infrastructure for NEVs and other low speed modes. LEAN networks are safe, dedicated routes with continuous flow features like small bridges, tunnels, and off-road paths. In addition to NEVs, these networks protect other low-cost alternatives to conventional vehicles like standing scooters, seated scooters, pedicabs, pedal-assist bicycles, bicycles, and all active modes.

These low speed, low emission modes can allow people to go the "first/last mile" to reach public transit stops. Solving the first/last mile problem is essential to making public transit more accessible to more people.

LEAN networks are a gentle disruptor. They are transformative, but no one loses. LEAN networks benefit virtually everyone in some way, even someone who hangs onto his Hummer. In fact, the greatest advocates for LEAN networks may well be the drivers of conventional vehicles. When they see an NEV driving past, I want them to think positive thoughts:

"That's one less car on the road." *Less congestion.*

"How great that people can get around so much easier and cheaper." *Social equity.*

"Now *that's* a smart way to help stop climate change." *Save our planet.*

"How great that we could increase the capacity of our roads without having to build a lot of expensive new roads." *Reduce taxes.*

Existing pedestrian and bicycle off-road infrastructure

Winding off-road neighborhood path

Shared bicycles and light rail – multimodal in Dallas

Iconic bridge near Austin Public Library

Innovators and Early Adopters

Neighborhood Electric Vehicles (NEVs) aren't new, but the idea that they could be adopted on a scale that could help relieve traffic congestion in cities and become a significant mobility alternative is new. Let's dive into what the innovator and early adopter cities and regions are doing.

1. **Lincoln, California** (pop. 47,000) was the first city in the U.S. to incorporate NEVs into its mobility plans. Its approach has become a model for small cities around the world with visitors most frequently from California and Arizona and from as far away as China, Japan, and Korea. What they find is that Lincoln is continually assessing, extending, and improving its NEV network.

    The movement began with golf carts owned by residents of some of the 6,500 homes in Sun City Lincoln Hills, a retirement community with two championship golf courses. Golf carts began to spill over on sidewalks and city streets, and rather than fight or ignore this, in 2005 the City of Lincoln recognized a need. Its City Council requested a city-wide NEV network plan that would enable any resident to travel from home to downtown Lincoln. Lincoln began educating its residents on the difference between NEVs and golf carts. NEVs' safer design allows them to travel on city streets, where golf carts can't go. Now as golf carts age out, most of their owners are spending the extra money to replace them with NEVs, which can also be used on golf courses. Charging stations are part of the standard for new real estate developments so people can charge their NEVs while they shop or dine out, extending their range. Lincoln now has approximately forty miles of multi-use lanes, striped but not physically separated from conventional traffic. This works for Lincoln because motorists are used to seeing NEVs. Lincoln is growing, and more young working professionals are moving in and starting to opt for an NEV instead of a second or third household car.

"NEVs are part of the culture of the City of Lincoln. They're part of the vernacular," said Ray Leftwich, City Engineer for the City of Lincoln. You can find Lincoln's NEV plan on the city's website.

2. **Peachtree City, Georgia** is a golf course-oriented community near Atlanta, Georgia. Over the past decade, golf carts multiplied organically, with kids driving them to high school. The city now has one hundred miles of off-road paths for golf carts, NEVs, and other low speed modes. The paths are separated from conventional traffic.

Peachtree City also demonstrates how little protected networks can cost. It recently added twenty-six miles to its network of off-road paths for $10 million. That's roughly $400,000 per mile. Peachtree City proves that LEAN networks don't need mega-million-dollar, multi-ton concrete bridges to achieve continuous flow. Its wooden bridges work fine, and they're charming. The paths also have small tunnels fitted into slopes with vegetation. Peachtree City's low speed paths encourage place-making along the way. The network of paths is the city's hallmark. TripAdvisor lists the paths as the #1 thing to do when visiting the city. You can find the city's Multi-Use Path Master Plan online.

3. **Coachella Valley, California** is developing a fifty-mile transportation and recreation pathway connecting cities and lands of three tribes. The first two-mile segment opened in 2018. Called CV Link, the pathway follows the Whitewater River flood channel between Palm Springs and Coachella. CV Link parallels Highway 111, the busiest corridor in the valley. CV Link has dual paths with grade separation for pedestrians, bicyclists, golf carts, and NEVs. One of the key goals is to reduce deadly conflicts with conventional vehicles by separating these modes.

   The award-winning project is led by the Coachella Valley Association of Governments (CVAG), with its thirteen member cities and tribes. The $100 million cost works out to about $2 million per mile. What I am most struck by is CVAG's vision of CV Link as a spine that enables cities and tribes along the way to develop low speed networks that connect to other cities though to CV

Link. San Diego Association of Governments (SANDAG) offers a webinar featuring transportation engineers from Lincoln and Coachella Valley discussing their networks.

4. **LA Metro (Los Angeles)** has developed a "Slow Speed Network Strategic Plan for The South Bay." Its approach is a transition that buffers pedestrians and bicyclists with low speed motorized modes including NEVs. LA Metro is taking somewhat of a risk, in my view, by not protecting NEVs more. But it is a visionary start and a model for cities in nonattainment ranging from serious to extreme. For you, LA Metro's plan contains excellent information, artist renderings, and schematics. You will see how the plan complements Complete Streets with "a more fine-grained, high-resolution approach to providing an infrastructure for the widest possible range of vehicles that travel at or less than twenty-five mph." You will see how it benefits pedestrians by providing a buffer that "disrupts the hegemony of the car." It is the most futuristic of all the plans that I've seen. In creating a regional network of "Slow Mode Thruways," LA Metro's plan anticipates that there will be much more to come in mobility innovation. You can find the plan online.

You'll notice that the first two cities began with golf carts since they were already present. But for cities that are developing LEAN networks from scratch, I don't recommend including golf carts. They aren't designed for city streets. They aren't even that safe on golf courses. The most common golf cart injuries occur when people simply fall out or the cart tips over. The American Academy of Pediatrics recommends that children not be allowed in golf carts because of the risk of injury. NEVs cost more, but the cost difference is roughly equivalent to a trip to the emergency room. People see how fun and easy golf carts are, but they don't realize the risk of injury. So it's safest not to include them in LEAN networks except within golf communities.

These innovator cities – Lincoln, Coachella Valley, Peachtree City, Los Angeles – have hit upon something extraordinary: a low cost, desirable mobility alternative that that can help achieve a region's air quality goals. This positive solution seems more obvious with every passing day.

LEAN Networks for the Birds

When I first began talking with people about protected low speed networks as a mobility alternative, people were intrigued. Often someone would ask, "What if we build it, and they don't come?"

It's a fair question. Which came first, the chicken or the egg? To answer it, I pointed to places where people were already using NEVs, like Lincoln, California; Bay Harbor, Michigan; and Peachtree City, Georgia.

But then something phenomenal happened. The "Birds" landed.

On April 15, 2018, I noticed a flock of young men on scooters zipping through downtown Austin. I instantly turned in their direction. I learned they were in Austin for a friend's wedding and were exploring the city on dockless scooters.

Within a matter of weeks, dockless modes – from companies like Bird, Lime, and Jump – landed in cities across the country. Transportation researchers call their adoption rates "unprecedented." Others have called it "an invasion, seemingly overnight." In one swoop, they disrupted mobility in major U.S. cities, accelerating the move to low-carbon-emission modes by demonstrating demand for short-distance, low-cost shared mobility.

In the first six months in Austin, there were 1.2 million scooter riders. At this writing, there are more than 300,000 scooter trips per month. That's 10,000 trips per day. Other cities are reporting similarly explosive demand. City of Dallas Director of Transportation Michael Rogers describes how rapidly technology is changing mobility.

> Take a look at how quickly scooters and bikes have made a difference in Dallas and in other cities across the country. Who would have thought? It's been less than a year and some of the numbers are astounding – hundreds of thousands of trips being taken. And those trips are during the peak hours when you have the most congestion of vehicles on the streets. So what this is doing is taking a vehicle off the street. It's an opportunity to get our area closer to nonattainment every time you see an electric vehicle or a self-propelled vehicle on the street... With the issues of congestion, the environment, and safety, I see the emergent technology is coming really quickly, and it's here to stay. It's making a difference, and it's going to make a greater difference.
>
> Our municipal governments, our county, federal, and state governments, have to make ourselves ready for these technologies. They're coming so quickly that we don't even have a way to look at the regulations. We were in the process of trying to regulate dockless bicycles when the scooters arrived. That's how quickly the technology is moving forward. So governments need to be swifter and more innovative in allowing more of these innovative ideas and practices to be

integrated into our formal operations because these modes of transportation are going to be normalized in our world.

In Austin, the rapid adoption of scooters put to rest the chicken-or-the-egg question for LEAN networks. Since the Birds landed and won over riders, that question is moot. It's now proven that people will use flexible, affordable mobility if it's available. Cities everywhere now recognize that the new dockless modes are here to stay, like them or not. Go to smartcitiesdive.com to find "Smart Cities Dive" and its interactive map of U.S. cities where dockless scooters and bikes are operating.

As scooter injuries mount, cities are also learning about the dangers. Weaving while sharing roads and sidewalks, ignoring traffic rules, and coming into conflict with cars and pedestrians is ultimately fraught with danger. We need to transition to better ways to protect these popular modes.

Austin has one of the largest deployments of dockless scooters in the country. In response to injuries, the City's Transportation Department led by Rob Spillar reached out to the Centers for Disease Control (CDC) to study the health risks. The CDC is examining incidents and injuries that occurred in Austin over a sixty-day period. The findings should help cities develop safer practices.

It's clear now that LEAN networks aren't some crazy idea. More U.S. cities are including NEVs in their mobility plans. Others are working on it. Scooters have forced the issue for many cities, but it's not happening fast enough to save our planet yet.

That's where you come in. This book is your guide to leading your member cities to improve mobility while dramatically reducing carbon emissions. I will walk you through how to introduce LEAN networks to your cities so that people are not only receptive to change, they are energized, enthused, and on board.

Chapter 4:

Meet Neighborhood Electric Vehicles (NEVs)

Now let me introduce you to Neighborhood Electric Vehicles (NEVs) so that you can introduce them to the people in your cities.

NEVs have a maximum speed of twenty-five mph. The U.S. Department of Transportation calls them Low Speed Vehicles (LSVs). We choose to use the friendlier term Neighborhood Electric Vehicles since virtually all LSVs are electric. If you hear the term "LSV," just know that NEV and LSV are interchangeable terms.

The U.S. market leader in NEVs is Polaris, the Minnesota-based company best known for its suite of specialty vehicles – like all-terrain vehicles, snowmobiles, and light haulers. To date, more than 55,000 Polaris GEMs have been sold. Here are several configurations.

Polaris GEM's versatile eL XD

Polaris GEM's human scale e4

Polaris GEM's e4 holds four people

Polaris GEM's two seat utility vehicle

NEVs are exempt from federal safety standards required of conventional vehicles. Being small and lightweight, NEVs don't offer passengers the protection and safety of conventional vehicles. This is why most states restrict them from roads where the speed limit is thirty-five or forty-five mph, depending on the state. NEVs' vulnerability is why I believe LEAN networks are necessary. It is essential that NEVs be protected from conflicts with conventional vehicles. Like bicycles, scooters, pedicabs, Segways, or any low speed mode, they will lose in a crash with a full-size conventional vehicle.

NEVs aren't new. They're proven. You can find them in use all over the U.S. and around the globe. NEVs come in all types – two seats, four seats, six-seat shuttles, pickups, and utility trucks.

Their versatility makes NEVs suitable for all kinds of purposes – errands, city centers, work trucks, landscape maintenance, security patrols, military bases, shuttles, even food trucks and ice cream trucks. Many are open-air. Others have detachable doors. Some have air conditioning. You're most likely to see NEVs as part of a fleet on a corporate or college campus or driving around planned use communities in such uses as:

- Security patrols at malls
- Landscaping trucks in parks and on campuses
- Maintenance vehicles at apartment complexes
- Police patrols in cities
- Resorts, both for residents and property management
- Golf communities
- Active communities for people over age fifty-five

NEVs are lightweight and easy to maneuver. Most weigh between 1,200 and 2,500 pounds, about a third of what conventional cars weigh. They have a tight turning radius and require little space to park. Most are 4'8" wide or less. Almost everything about them is smaller – except that they have surprisingly ample legroom and headroom even for people as tall as 6'8". They're much easier to get in and out of than conventional cars.

You can charge an NEV at any 110-outlet, no different than your coffee-maker. You don't need a special charging system. Their range is typically thirty miles – easily accommodating most errands or short commutes. Owners can add batteries to extend the range.

By law, NEVs have these features: auto safety windshield glass, three-point seat belts, running lights, headlights, brake lights, a horn, reflectors, rear view mirrors, turn signals, windshield wipers, and on-road tires – features not required on golf carts.

One Quarter of the Cost

The purchase price of a new Polaris GEM is $10,000 and up. A six-seat shuttle will run $16,000 or more.

But the cost advantage is not just the purchase price. It's the cost of owning, maintaining, and powering an NEV. *Over its life, an NEV is roughly one-quarter of the cost of owning, operating, and maintaining a conventional vehicle.*

The day you start driving an NEV, you start saving money. NEV insurance is typically $200/year or less. Electricity costs about $100/year. Maintenance is minimal since NEVs have clean electric engines. There is an array of battery options at different price points. The largest maintenance expense is replacing the battery.

Compare this to the true cost of owning a conventional car. For a typical car driven 15,000 miles/year, owning, operating, and maintaining that car averages $8,470/year, according to the American Automobile Association (AAA). That number includes car payment, depreciation, gas, insurance, maintenance and repairs, and registration. That's over $700/month. If a family's net income after taxes is $35,000, that car takes 24 percent of their income.

In comparison, an NEV can cost less than $2,400/year, a monthly cost of $200. Of course, an NEV will not be driven on long trips, so its annual miles driven will be far less.

But compare a monthly cost of $700 to $200. The difference is $500/month.

That $500/month can be the difference between sacrificing necessities and being financially sound. More than that, it can mean independent mobility for people who otherwise could not afford a vehicle. It can mean getting to public transit to get to a job or eliminating a transfer on the bus and getting home in time to make dinner for the kids.

Eliminate Carbon Emissions

NEVs use electricity. The typical NEV generates about 100 to 150 kilograms of CO_2/year from its electrical source, typically a power plant.

In comparison, a conventional gas-powered car emits *4.6 metric tons* of carbon dioxide (CO_2) per year. That is nearly one hundred times more CO_2 than an NEV.

Consider this. If seventy percent of a fleet of conventional combustion vehicles were converted to Polaris GEM's all electric NEVs, it would eliminate 3.7 million pounds of CO_2 output. It would also save $1.5 million over seven years.

Compare Carbon Emissions

A conventional vehicle emits 4.6 metric tons of CO2 each year while an NEV emits a tiny fraction of that

For every one hundred combustion vehicles replaced with Polaris GEM's all electric NEVs, 2,100 trees would be saved.

To dig into the data, search online for the EPA's Fast Facts on Transportation Greenhouse Gas Emissions and Carbon Footprint Calculator.

It is clear. Transitioning to NEVs is a game changer for carbon emissions.

NEV specs

I use NEV specifications to define requirements of LEAN networks. Ideally:

- LEAN Lanes should be ten feet wide for two-way traffic.
- LEAN Lanes and bridges should be designed to handle vehicles that are a maximum of 2,500 lbs.
- LEAN Lanes' maximum speed limit is twenty-five mph. NEVs are designed not to exceed twenty-five mph. The federal definition of low speed is a maximum of twenty-five mph.
- LEAN features such as spiral ramps should be designed to accommodate NEVs with lengths up to that of a six-seat NEV shuttle.
- Curves in LEAN Lanes should be designed to handle an NEV at certain speeds, e.g., an NEV may need to slow down to ten mph for a tight curve due to terrain or right-of-way, and this should be posted.

Another reason that I focus on NEVs is that they are most like the conventional cars, SUVs, and trucks that we use today. They are the most comfortable. You know what to expect: that you will buy it, own it, park it, maintain it, and insure it, just as you would any car, except that it would be at a fraction of the price and you would recharge it instead of refilling with gasoline. This is the easiest transition for someone who doesn't want to give up the convenience and comfort of a car.

Electric Cab North America

So why aren't we driving NEVs on short trips? What's the hold-up?

New paradigms are hard, especially when existing paradigms are codified in laws, policies, and regulation. Changing things can take years of effort to overcome daunting obstacles.

Take the story of Electric Cab North America, whose mantra is "urban mobility on demand." Founder Chris Nielsen's battle began in 2008 when he petitioned the City of Austin to allow his transportation business to operate as an urban mobility service. The City responded by issuing his company 227 citations, leading to multiple court appearances. He won nine separate recommendations from the Urban Transportation Commission to create a license for low speed vehicles so that a company like his could operate commercially. The recommendations were ignored. After five grueling years, the City Council finally relented and granted Electric Cab an Operating Authority but at a great personal cost to the founders.

In spite of this, Electric Cab has grown to be the largest fleet operator of its kind in the U.S.

Now, of course, Electric Cab is established and valued. Its riders don't know that there was ever such a struggle. They just know that Electric Cab will easily shuttle them around downtown, so they don't need a car. Those who once opposed Electric Cab now appreciate and credit what it does for the city. They speak with pride about the innovative national company based in Austin. I admire the City Council and staff who relented because it required one of the most difficult of human endeavors – thinking differently.

I relay this story because it's important to know that "business as usual" may not comprehend that there is a transformation underway in urban mobility.

Electric Cab North America, based in Austin, Texas

Pedicabs

It doesn't stop with NEVs. There are a number of other types of low speed, low emission, low impact vehicles that are just as efficient, affordable, and available in bridging first/last mile transportation gaps.

My nominee for the most fun, experiential alternative is today's modern pedicab. David Knipp and Ken Cameron own and operate Movemint Bike Cab in Austin. Together they manage a 60+ pedicab fleet in the heart of the city.

Over the last several years, Movemint has not only built one of the nicest fleets in the country, they've also built a machine shop/manufacturing space where they design, build, and manufacture pedicabs, industry accessories, and electric motor drive systems. Movemint's manufacturing line, Precision Pedicab, has been called 'the Tesla of Pedicabs' because of its thoughtful design and David's talent for continually enhancing it.

In 2017, David and a couple of other industry enthusiasts worked diligently to get the City of Austin's attention and permission to entertain an Electric Assist Pedicab Pilot program. They succeeded, and pedicabs are now released from a non-motorized ordinance as a pilot project. Pedicabs have always been an efficient mode. Now they can perform beyond a driver's physical endurance alone. At a time when technology and mobility are evolving, pedicabs can evolve too. Electric assist allows drivers to go farther on Austin's hilly terrain. The pedicab industry is now a viable source of mobility solutions for distances up to two or three miles. Drivers can comfortably work longer hours. Electric assist takes passenger guilt out of the equation.

Ten years in, Movemint and Precision believe they're still just getting started. If you are interested in bringing pedicabs to your city, talk to David and Ken at AustinPedicab.org.

Movemint Bike Cab, the Tesla of pedicabs

Exploring downtown Austin by pedicab

Scoot and More

The company Scoot is based in San Francisco, where it has developed original, small, lightweight, bright red electric shared vehicles. Its first market is San Francisco, where it operates a fleet on city streets. Its focus is getting people out of their single occupancy cars by offering an easy mobility option that works for intra-city trips and makes multimodal connections easy.

Unlike scooter companies, Scoot worked with the City of San Francisco to develop standards and operate with the City's approval. Scoot is building a following of loyal users with such a sense of community that they have meetups.

Having ridden Scoot around San Francisco, I love that people smile and children wave when they see a Scoot go by. Scoot has a *Happy City* philosophy, that people are happiest when cities are human-scale and people-centered instead of auto-centric. Scoot is expanding in Barcelona, where there are car-free zones, and the city welcomes low speed and self-propelled shared vehicles.

Multimodal, app-based movement with Scoot

Scoot, based in San Francisco

Independent Mobility

All of this is to say that Tiny Transit is comprised of all kinds of vehicles – shared or owned; electric or self-propelled; enclosed or open air; two, three, or four wheels. What they have in common is low speed, low emission, independent mobility.

They give people options and choices. They allow people to save money while reducing their carbon footprints.

They free people from dependence on internal combustion engines and conventional size vehicles.

They free cities to spend less on transportation while improving mobility choices for everyone.

Whatever the style of vehicle, each ride has meaning for our air and the environment.

Chapter 5:

Think Vision Zero

The prime directive for LEAN networks is safety.

Zero fatalities. Zero life-altering injuries. Nothing is more important – not speed, not efficiency, not convenience, not wait times at intersections, not anyone's feelings, not anyone's resistance, and certainly not money.

Nothing.

What's the number two directive for LEAN networks? Safety. Number three? Safety. Instead of the real estate mantra "location, location, location," think *safety, safety, safety*.

That's it. There can be no compromise.

Safety is the first secret to transforming your member cities into models for low carbon emissions. Once you can demonstrate to your community that LEAN Lanes are safe, you will see an enormous upsurge of users. People will have the confidence to invest in NEVs, better scooters, pedicabs, and commuter bicycles with pedal-assist.

If you fail to do so, public opinion will be as merciless as it has been for fatalities in the first applications of driverless cars. Statistics show that self-driving cars are safer when you weigh millions of miles driven. But ordinary people don't drive millions of miles. They drive to work. One life lost is one life too many.

To achieve your goals for reducing carbon emissions, you need community buy-in. People will not buy in if users are getting injured or killed, and they shouldn't. You are doing this to *help people*, not get them killed.

Remember, cities will be asking something intensely personal. They will be asking parents to entrust their safety and the safety of their teenage drivers to the LEAN network. Nothing is more personal or sacred than that trust.

Changing the Paradigm

Believe it or not, one of the major points of resistance will likely be some of your transportation experts and engineers. The problem is, safety is lower down on their lists of what's important. They can't help it. It's how they were trained.

You might think, "Let's get the best transportation engineers involved to ensure safety. We'll make them responsible for ensuring the network is safe."

Wrong again.

Not long ago, I was meeting with a transportation planner. We were discussing incorporating LEAN networks adjacent to a highway. As I looked at one concept, I saw that it counted on all drivers being good drivers.

If there's one thing I'm certain of, it's that all drivers are *not* good drivers. A lot are terrible, and even the best drivers make mistakes. I hope I'm not just speaking for myself when I say we have all turned the wrong way onto a one-way street, gone forty-five mph on a thirty-mph road without realizing the speed limit, started to change lanes when suddenly, as if dropped by the heavens, there is a car in that space, or completely missed a red light – that is the most terrifying. When designing roads, we have to anticipate all of the mistakes drivers can make.

So I drew attention to that dangerous exposure in the plans. The transportation professional responded that the odds of getting hit were no worse than for a conventional car.

At moments like that, and there have been a few, I have to restrain myself! And I remind myself that this concept and our uncompromising approach to Vision Zero design are new to people.

From your position with a city, COG, or MPO, you must put safety front and center. Reiterate *safety, safety, safety* as the prime directive from the get-go. Don't let yourself be talked out of it as impractical or unachievable. Don't let someone with greater professional expertise convince you that Vision Zero is impossible.

LEAN networks must be designed with zero tolerance for fatalities and life-altering injuries. To accomplish this, you and the decision-makers in your cities must flatly reject any compromise or effort to water down public safety.

Twelve Ways to Protect Low Speed Modes

You might wonder what I mean by *protected* networks. With LEAN networks, we're talking about physically protecting drivers and passengers who are using low speed modes from collisions with conventional cars and trucks. If there is a collision, the vulnerable modes could be crushed, and their precious cargo could be hurt or killed.

If we do not address this exposure with a Vision Zero approach, we will lose the chance to win drivers over to this low emission alternative. Protection is essential. Here are twelve ways LEAN networks can protect low speed users.

1. ***Multi-use off-road paths.*** I list these first because they are ideal. A twelve-foot path provides enough space not only for NEVs at 4'8" or less, but also for other low speed modes. Peachtree City, Georgia's network of multi-use paths accommodates bicycles, NEVs, golf carts, and other low speed modes. The multi-use approach eases the transition to these new modes. Take Katie, for example. One day she bicycles to work, the next she'll take her NEV to the grocery store, and on the weekend, she may drive her daughter to a sports competition forty miles away. The key is understanding that users are all in this together, and it's in everyone's interest to share space. As the network is expanded and miles of off-road paths are developed, all modes benefit.

2. ***Moveable barriers.*** There are hundreds of thousands of concrete, water, or sand-filled barriers lying unused around the U.S. They're inexpensive and immediately available. These can be installed over the course of a weekend and can be easily moved if it turns out they're not in the right place. Motorists understand barriers. They're not new.

3. ***Ordinary bollards.*** You see these concrete poles everywhere once you start looking for them. Often you see them at building entrances to stop cars and trucks from plowing into buildings. They can be used to keep conventional vehicles off of off-road paths or to firmly define a LEAN

Lane once it has been proven to work well with moveable barriers.

4. *Retractable bollards.* You may not have seen these or noticed them before. At the University of Texas at Austin, lighted bollards rise from the pavement to stop traffic between classes to protect pedestrians and cyclists, and then voila! They retract into the ground once classes start. They disappear.

5. *Planters.* Large planters stop traffic as efficiently as bollards or barriers, and they are beautiful. These can be expensive to buy and maintain with plantings, but the expense makes sense for certain locations where they can become part of place-making or way-finding. For example, they can work in commercial, hotel, and retail locations as beautiful features with flair. A fancy hotel, for example, might well pay for planters.

6. *Small tunnels.* Tunnels needn't be dark and foreboding. They can be well-lit and attractive, even carved through embankments with vegetation. Peachtree City sets the standard with its charming tunnels.

7. *Bridges.* You may have read about pedestrian or bicycle bridges that cost $15 million or more. I am at a loss to understand these huge expenditures. It just seems like a staggering amount of money. Yes, you could pay that much, but it is necessary? Is it a wise expenditure? Maybe it is. Maybe not. Peachtree City has shown that it might be possible to build an additional thirty-six-plus miles of

off-road paths for that amount, including small tunnels and bridges along the way. Small bridges can be simple and cost as little as $60,000 to $70,000 because low speed modes are so lightweight.

8. *Bike/pedestrian bridges.* Not all, but some, existing bridges could be adapted to become multi-use. I understand that some of these bridges were hard-fought wins for the pedestrian and bicycle advocates, yet I believe that we are all on the same side of carbon emissions and safety, and that these bridges may help establish the viability and demand for a LEAN alternative. NEVs and other low speed motorized modes can be required to yield to bicyclists and pedestrians. There may be a need for a crossing signal that allows NEVs and other low speed modes to cross only when pedestrians are not present, for example.

9. *Metal guardrails.* You can say this for them – they work! I'm referring to both the full metal railings and those that are heavy-duty wires between wooden posts. The materials aren't expensive. They provide excellent protection for around $100,000 for the materials for one mile of guardrail.

10. *Grade separation.* This is also ideal, with complete physical separation for low speed modes.

11. *Barrier arms.* You see this classic solution at railroad and light rail crossings. Some barrier arms are stronger than they look, stopping cars almost as surely as concrete.

12. *Place-making, way-finding art and trees.* Bollards, planters, and landscaping can be surprisingly beautiful. They can even be public art.

Planters as beautiful barriers

Trees and bollards create an effective barrier

Railroad crossing arms as a barrier

Lighted bollards at dusk

Retractable bollard sign on UT Austin campus

Retractable bollard

You'll find a number of images of these kinds of barriers on TinyTransit.com, most of which you are free to use. Most are photos I've taken, and you have my permission.

You may notice that some of the mechanisms to protect low speed vehicles also help them move along. By incorporating bridges, tunnels, and smart designs so that the NEV drivers are flowing at twenty-five mph, they will pass drivers who are moving more slowly on congested roads. This goes to the key finding of

Katie's dissertation – that with continuous flow during rush hour, "you can get there faster by going slower." Not to mention safer.

Believing Change Is Possible

Believe that it is possible to eliminate traffic fatalities. If you use the term Vision Zero but in your heart you don't think it's possible, then that doubt will reflect in subtle ways in your work.

I compare this to the scene in a murder mystery where the police are interviewing the husband about the disappearance of his wife. He's doing swimmingly until he refers to his wife in the past tense. He is revealed.

In such unconscious ways, you will reveal that you don't believe Vision Zero is possible. So, believe it. Own it. This story from my experience might help.

Years ago, a stretch along Highway 183 in Austin had been the scene of many fatalities. It was so famously dangerous that there were bumper stickers: "Pray for me, I drive 183." By comparison, fatalities on the newly opened Loop 1 were rare – a single fatality over the course of that year. At the time, I worked as the aide to City Council Member Lee Cooke, who as personnel director for Texas Instruments, located on 183, knew several of those killed. Lee refused to accept transportation professionals' assessment that nothing could be done or that it would take years. He enlisted the City Manager in imposing a six-month moratorium on new development along that deadly corridor while accelerating roadway improvements, stepping up enforcement, and other measures. Developers hated the moratorium. It was the last thing they'd expected from a business-oriented council member, but almost immediately, fatalities dropped. Crashes and their severity declined.

What I learned from this was that the problem was not the drivers. The problem was roadway design. I learned that to refer to a collision as an "accident" is to dismiss the power we have to change things. It's loose talk that lets a defective system off the hook. I learned that there are often perverse trade-offs between money and lives, between lazy thinking and lives. I learned that resolve grows when someone like Lee must break the news to a young family that their father was killed.

Cautionary Tale

In 2003, the National Transportation Safety Administration (NTSA) had evidence of danger posed by drivers using cell phones while at the wheel. There were hundreds of pages of research and warnings, all withheld from the public for six years. In 2009, the materials were made public by safety advocates through a Freedom of Information Act lawsuit. We learned that the former head of NTSA was "urged to withhold the research to avoid antagonizing members of Congress." The warnings were buried, including this: "We therefore recommend that that drivers not use wireless communication devices, including text messaging systems, when driving, except in an emergency." And this: "[We] have concluded that the use of cellphones while driving has contributed to an increasing number of crashes, injuries, and fatalities."

Who in good conscience would have suppressed these research findings? We may never know how much influence the cellphone industry had since advocates for transparency were not "in the room where it happened." We do know that the former head of the NTSA withheld the research to avoid antagonizing the House Appropriations Committee, jeopardizing billions in appropriations. "My advisers upstairs said we should not poke a finger in the eye of the Appropriations Committee," he said years later.

During the years that followed, tens of thousands of lives were lost and a pervasive culture of multitasking while driving took hold across America.

Finally, in 2009, the 2002 study was released.

Looking back, the decision to withhold research was unconscionable. The genie escaped the bottle and has wreaked havoc across the U.S. ever since.

The lesson of that disturbing story is that now is the time to get it right. Cities need to get ahead of this new mobility by providing for its safe passage. It goes back to that sacred trust.

Once safety is compromised, cities have lost that trust. Decide that you will not wait until tragedy strikes to take the actions necessary to preventing fatalities and serious injuries.

You have to believe in your heart and mind that Vision Zero is achievable on LEAN networks.

Science and Psychology

A pedestrian hit by a vehicle going forty-five mph has a ninety-five percent chance of being killed. But if the vehicle is going twenty mph, the chance of being killed is five percent. Slower is clearly safer.

And there's another reason low speeds are safer. Consider how speed affects peripheral vision. This graphic compares what drivers see at fifteen mph with their field of vision at thirty mph. Driving more slowly not only reduces the impact of collisions, it also reduces the *potential risk* of collisions. Going slower dramatically improves drivers' ability to see what is around them.

Speed reduces drivers' vision and ability to see pedestrians.
Source: Local Government Commission

Director Werner Hertzog's 35-minute film, "From One Second to the Next" presents four stories of lives shattered by drivers texting on their phones. It reminds us how very precarious our lives are when drivers' minds are focused somewhere else other than on the road. Find the film on YouTube.

What more can we learn from psychology and behavioral science? You may know the famous "invisible gorilla," one of the best-known experiments in psychology.

"The experiment reveals two things," says Daniel Simons, who developed it. "That we are missing a lot of what goes on around us, and that we have no idea that we are missing so much." Find by searching "the invisible gorilla."

We cannot assume that people are continually mindful of their surroundings. Even without distractions, even with intent focus, people miss much of what is around them.

This mindlessness is why I urge that the U.S. Department of Transportation revise its safety standards for NEVs to require that they are in neon colors and have low speed triangles on the rear. At present, drivers are not looking for NEVs, just as video viewers weren't looking for a gorilla. We cannot have NEVs become "invisible gorillas" on our roads.

Tyler, Part Two

In the days after Tyler's injury, I went to look at the crosswalk where he was struck. He and I had bicycled through it before, but I'd never noticed what I now saw. It was a midblock crosswalk without any lights or signals, without a button to press for a light to stop oncoming traffic. The reflective stripes on the pavement were worn away. The signs were pathetic. It was on Barton Springs Road, which might as well be called Margarita Row. All along the road were commercial establishments with their own large, attention-grabbing signs. It was a few yards from a restaurant known for its wacky decor. How boring it is to watch for pedestrian signs when a hubcap collage and Elvis memorial are competing for your attention.

I asked for a meeting with the City transportation department to see what could be done to improve safety at that crosswalk. To get the meeting, I assured them that I would not sue the City.

The staff were very sympathetic. I learned that a midblock, unsignaled crosswalk is called an "uncontrolled crosswalk," and that what happened to Tyler was a "triple threat." I was told that to upgrade the crosswalk would cost $40,000 or more. Bear in mind that Tyler's medical bills exceeded $20,000. The City's transportation staff agreed to look at the intersection but determined that it didn't require an upgrade to a signal. They did refresh the faded paint on the asphalt and post new signs.

I learned that the City didn't know how many "uncontrolled crosswalks" existed in the city or where they were. There was an overall transportation inventory underway.

I am not picking on these dedicated staff. They were doing their best with the tools they had, no different than most cities across the country.

I talked with an engineer who told me that he has a strategy for proceeding through uncontrolled crosswalks. He makes sure that he catches the eye of the first driver, then pauses for the driver in the next lane, catches his eye, and proceeds. This is all well and good for an engineer in his thirties, but how is a fourteen-year-old to have enough crosswalk experience to develop a strategy? It is too much to expect of a child.

I wanted to see the criteria used to determine that the crosswalk was safe. I was referred to a thick report with a lot of words and math. It seemed as if I was discovering the awful truth behind road safety. It isn't whether a road or crosswalk is safe or not. Clearly, the Barton Springs Road crosswalk was unsafe. Instead, what transportation professionals look for is, does it meet the criteria set forth by the transportation industry? If it does, the professionals are protected from liability if someone is injured or killed.

Attempting to educate drivers is not sufficient. How exactly are we going to educate drivers to come to a complete stop at midblock, unsignaled crossings just in case there is a pedestrian or cyclist hidden from their view? This relies way too much on drivers to be focused at all times, courteous, careful, alert, and educated on the dangerous phenomenon of "uncontrolled crosswalks." I study traffic safety, and I didn't even know about this crosswalk offshoot until Tyler was struck in one

In investigating Tyler's collision, I learned firsthand the perverse tradeoffs transportation engineers must make. For the sake of driver convenience, their guidelines compromise safety. They make trade-offs that I suspect you or I would not make. The trade-offs are baked into roadway design standards. Our roads today reflect decades of such design. It shouldn't take a dead child to prove the need to upgrade an indisputably dangerous crosswalk. Tyler's near-fatal experience solidified my commitment to developing a Vision Zero mobility alternative.

Hunger Games Tributes

All of this is to say that the dangers on our roads should terrify you. We have accepted the Hunger Games model where we sacrifice 35,000 lives each year to allow the rest of us to drive. The lack of will, lack of resolve, and lack of urgency are unacceptable.

The ugliness is only getting worse as affluent drivers can afford new vehicles with expensive safety features while low income people are disproportionately killed. Cities in Texas are seeing an alarming increase in pedestrian and bicyclist fatalities. Incident reports indicate that a disproportionate number of pedestrians killed are people of color, people with low incomes, and people who do not own cars. Their names are in the "reaping" several times so their odds of being chosen are worse. Even so, no one is safe.

Shouldn't we demand as high a standard from our transportation system as we do from our cell phone providers? If cell phones killed 40,000 Americans every year, don't you think we would be outraged?

To succeed, LEAN networks must meet a higher safety standard. We are introducing new lighter-weight modes that cannot protect people sufficiently in the event of a collision with a two-ton conventional vehicle at forty-five mph. This is why the LEAN design must be different.

Creativity, Ingenuity

Vision Zero takes more than money. It takes creativity, ingenuity, and a touch of genius. Pretend your challenge is to create more window space on one side of an office building than it presently has – and it's already all glass. How is that even possible? Think of angling walls in and out. With their creativity, architects, engineers, scientists, software designers, and all kinds of professionals meet design challenges like this every day. If we can pull down the silo walls around transportation, we can enlist some of our best thinkers in solving LEAN network design challenges.

I am just as confident that there is enough genius in communities all over the U.S. to meet each locational challenge of LEAN networks. Our nation put a man on the moon and got him back. We ought to be able to get an American to work and back consistently without serious injury.

There are many transportation professionals who care and will help you. They are just as dismayed at our society's and their profession's tolerance for roadway fatalities as I am. You don't need to look too hard for them. Once you bring this concept to your member cities, they will likely surface to help.

For your region to shift toward slower, low emission modes, the LEAN infrastructure must win *the public's sacred trust*. LEAN designers must anticipate and eliminate all permutations of conflicts between conventional vehicles and small, vulnerable modes.

You can convey this. For your cities, everything rides on this.

I'll end as I began. The first secret to transforming your community into a model for reducing carbon emissions is *safety, safety, safety*.

Chapter 6:

Build LEAN Networks with Crumbs

I t might surprise you to learn that the LEAN infrastructure can be built with the crumbs of what major transportation projects cost. Just think about what those cost:

- $8 million to $10 million for one mile of new four-lane highway in an urban area
- $3 million to $5 million for one mile of a new two-lane undivided road in an urban area
- $150 million for one new mile of light rail
- $150 million for one new mile of subway
- $480,000 for one mile of new sidewalk.
- $250,000 to $500,000 to install a new traffic signal

It's staggering, and aside from advances here and there, the U.S. has been doing it essentially the same way for over sixty years. A cloverleaf today is barely distinguishable from one built in the 1960s. An entire industry is built around designing and constructing major transportation infrastructure projects with behemoth budgets.

Transportation professionals are used to these costs. The people who manage these projects have the job of staying within budget. But the budgets are gargantuan, and the professionals don't have much incentive to reduce costs. It's almost the *opposite* incentive – secure a larger budget than is actually needed to be sure you don't exceed it.

If we approach LEAN networks in the same way as conventional transportation, they will be so costly that they will be too little, too late, if they happen at all.

Instead, I will show you how to take a *cost-efficient approach* to building the LEAN infrastructure. With practice, you'll learn to develop an eye for spotting underused assets, new developments *before* they're designed, and LEAN routes your community can't afford *not* to do. You'll banish any "woe is me" thinking that blocks creativity because you will begin to see opportunities all around you.

Cities Can't Afford *Not* to Create LEAN Networks

Your goal is to make identifying opportunities fun – so fun that people in your community will be drawn to the idea and want to help. Wait, you may be thinking… Can *infrastructure* really be *fun*? Yes, but you may need to retrain your brain.

I have sat through some of the most boring meetings ever. I get jumpy – getting up to get water when I'm not thirsty, passing notes, exquisitely doodling, stretching against the back wall. I'm sure you've been there.

Among the worst was a transportation commission meeting where I felt truly sorry for the commissioners. They seemed like bright, interesting, dedicated people, but they had been dragged into the transportation silo, their eyes dulled by PowerPoints.

After one meeting, I approached the chair to talk about LEAN infrastructure. I followed her all the way to her car. Finally, she turned and said, "Susan, when I hear you talk, I see dollar signs. The city doesn't have the money to do what you want. It can't even afford to build sidewalks. There's no money."

I realized I had not been communicating well.

That night, I started writing "How LEAN Networks Can Be Built with the Crumbs of Major Transportation Projects." That report is the foundation for this chapter.

The message is, "Communities can't afford *not* to build LEAN networks." I wrote it to break down silo walls. By using Austin as an example, it encourages readers everywhere to *participate* in the hunt to find smart, low cost ways to incorporate LEAN networks in their communities. There is room for everyone at the table.

By the end of this chapter, you'll learn how to turn the crumbs from major projects into an *advantage*. If you do it right – and I'll show you how – limited budgets can become community challenges that bring together diverse people to chime in and even jump into design process. A career transportation introvert might think, "No good can come of this!" But for you as you seek ways to reduce carbon emissions, this public involvement is invaluable.

How to achieve low costs – even ridiculously low – is the focus of this chapter. Let's walk through four types of cost-saving ideas that you can apply in your community.

1. Leverage existing assets
2. Find projects that are already funded
3. Incorporate LEAN Lanes into new developments
4. Right-size the network design

1. **Leverage existing assets**. To leverage assets, you need to look at your community with an explorer's eye, alert to spot assets that you could redeploy. Here are a few examples to spur your thinking.

   - **Baltimore City's underutilized roads**. Unused asphalt accrued as the city's population declined. My conservative estimate is $60+ million in downtown alone. This asset could be the foundation for a hub-and-spoke LEAN network serving all of Baltimore. And it would complement the exciting revitalization and resurgence of this historic American city.

   - **Underutilized highway frontage roads**. Every day I drive on a three-lane frontage road that would be amply served with two lanes, while converting the third lane into protected two-way LEAN Lanes.

   - **Off-road paths**. Earlier, I described what Peachtree City has accomplished in creating nearly one hundred miles of multi-use off-road paths for low speed vehicles and golf carts. They are spending about $400,000 per mile – a lot to you and me but mere crumbs in the high dollar world of transportation infrastructure.

   - **Share bicycle infrastructure**. There is a twelve-foot-wide bicycle bridge across an Austin highway, built at a cost of $16 million. It's wide enough to become a multi-use path for NEVs and other low speed modes without detracting from bicyclists. That creates an immediate $16 million asset. Will the bicyclists be willing to share? That is the challenge. I think that they will as they come to understand that bicyclists will benefit greatly with the introduction of slower modes, that the protected networks will protect them too, and that the transition cannot occur unless it is multiuse and multimodal.

- **Convert highway medians**. Along that same highway is a grassy median that extends for ten miles. It is a natural for converting to LEAN infrastructure for neighborhoods and destinations along the route.
- **Cantilevered bridges**. There are existing bridges along highways could be cantilevered to add LEAN Lanes. That vertical space is an asset.
- **Roads with excess road capacity**. Even in cities choked by congestion, there are roads with capacity to spare. In Austin, South Pleasant Valley Road is a prime example. It's six lanes wide, lightly traveled with expansive shoulders, just waiting for LEAN Lanes.
- **Unusual streets**. Every city has a quirky street that could incorporate two-way LEAN Lanes. In Austin, it's the peculiar Wallingwood Drive near Zilker Park. All three lanes are one-way. One of its three lanes could easily be converted to two LEAN Lanes simply by lining up barriers.
- **Connecting with transit**. A LEAN network could enable people to go the first/last mile to a transit stop, or even eliminate a transfer on the way to and from work, making buses, light rail, and subway lines more appealing and convenient. In effect, the transit agency is an ally, and the transit system is an asset. The LEAN network leverages its availability, increasing ridership and benefiting everyone.
- **Scouting locations**. One of David Knipp's colleagues and I have explored downtown Austin by pedicab. We've identified a possible four block LEAN route

that extends from a transit center all the way to the edge of Lady Bird Lake. Only one intersection on this route needs treatment, and that's to protect low speed vehicles as they cross a five-lane road. The intersection will be upgraded soon anyway; why not include a protected crossing?

Existing pedestrian/bicycle bridge across a highway

*Pedestrians on this bridge over a highway
might feel safer sharing it with NEVs*

Existing spiral ramp at pedestrian/bicycle bridge

Existing pedestrian/bicycle bridge over Houston's Gulf Freeway

*$16 million existing pedestrian/bicycle bridge
over a congested highway*

$16 million bridge is an asset

2. **Find projects that are already funded.**

Look for transportation projects in line item budgets for cities and counties where officials might be open to redirecting or expanding the projects to include LEAN Lanes. Think of it as a scavenger hunt.

We've examined bond language (what voters approve in bond referendums) and found that there is nothing in the language we've seen to prevent incorporating LEAN infrastructure.

These projects are already on the drawing boards. They are already funded. They are just missing LEAN Lanes.

As I write this, Katie Kam and I are developing a mobility plan for Community First! Village, an innovative community of tiny homes for people who are formerly homeless and disabled, located just outside Austin. Few residents have cars, and most lack drivers' licenses. We're working to redirect and expand a $2 million Travis County project to build sidewalk and bike lanes to include the LEAN infrastructure so that it offers independent mobility for everyone who lives and works at the Village. This will allow residents to get to jobs, transit, grocery stores, and classes on their own schedules. It will provide independent mobility.

3. **Incorporate LEAN Lanes into new developments.**

For the LEAN approach to work, it must help new developments become more marketable while containing or reducing development costs.

You can start by introducing yourself to city planners who know what developments are being proposed so you can reach out to them while they're still on the drawing board, before they're set in stone. Skim your local business journal to learn about large tracts of land changing hands and projects in the planning stages. Here are examples to prompt your thinking:

- A large, mixed-use infill development
- A new suburban residential development
- A new corporate campus
- A community college campus expansion
- A public/private downtown revitalization project that has financial backing
- Developments adjacent to a new transit center
- Smart corridors where city planners want to encourage more density

This approach doesn't work unless developers and their end users *want* to get on board. So give developers reasons to want to include LEAN infrastructure.

Imagine that a developer is allowed, encouraged, and incentivized to incorporate LEAN Lanes in the development and to build the network infrastructure linkages like bridges, flyovers, and tunnels around the development. This not only makes his development more appealing to buyers, it should reduce costs (e.g., think small footprint parking garages; garages designed to allow conversion) and reduce congestion involving conventional vehicles.

The key is, LEAN Lanes cannot be "in addition to" or they will simply add to the cost of the development, and a key benefit – reducing transportation's impact on the environment – would be defeated. For LEAN networks to work *for* developers, they must be allowed to *not* build certain roads, widths, and parking.

- Ask yourself, how can I make LEAN Lanes work *for developers?*
- Who needs a sea of parking that sits largely empty except on Black Friday? Ease up on parking space requirements and instead allow developers to incorporate LEAN infrastructure into the development, reducing parking requirements, and connecting destinations with the LEAN network.
- If there's a proposed development that has nearby neighborhoods protesting because of increased traffic – a common objection to developments – see if your city can help strike a compromise that mitigates traffic and incorporates LEAN Lanes along arterials, serving both the development and neighborhoods.

Your goal is to persuade the developer to include the LEAN infrastructure and persuade your city's mayor, city council, and planning and development office to make compromises on other requirements – say, parking, setbacks, fees – to incentivize the developer.

You also need those people to expedite the process for developers because they won't undertake anything new that could drag out the process.

If you wait too long, the developer will be so far into the project – having invested in zoning, site plans, permits, transportation plans – that it becomes impractical to redirect it.

Any new development in your community that does *not* include LEAN infrastructure is a missed opportunity that will not reappear. Once the built environment is built, you miss the window of opportunity, the only way to include a LEAN network may be an expensive retrofit at taxpayer expense instead of the developer's expense. It's far better for taxpayers to intercept and redirect these projects. Opportunity lost is opportunity cost.

Bridge over waterway in Silicon Valley

Utilitarian configurations are popping up, meeting needs

4. **Right-size the network design for these low speed, lightweight modes.**

You might be surprised at the low costs inherent in a LEAN network.

- **Charging**. Many cities are now constructing an electric charging infrastructure for conventional cars; that's great. But you can recharge NEVs and scooters at any 110 V outlet. The outlet you use to recharge your phone? It works for NEVs, too. Low speed modes use very little electricity (~$100/year for NEVs, less for scooters and other modes). They don't require a large-scale, expensive charging infrastructure. Of course, I'd like to see fast-charging stations for NEVs conveniently located along the way, but they're not essential, and their absence shouldn't hold you up.

- **Bridges and tunnels**. Low speed modes don't need mega-million-dollar, multi-ton concrete bridges to achieve continuous flow. As Peachtree City demonstrates, a $60,000 wooden bridge works fine. Peachtree City also has tunnels with an open feeling fitted into slopes with vegetation. It's all charming place-making.

- **Bollards and crossways**. Bollards are an NEV's best friend. *Retractable* bollards are posts that rise from the pavement to let NEVs safely cross streets while cross traffic is stopped. Placed four to five feet apart, they can protect NEVs through intersections where tunnels, bridges, or rerouting aren't possible.

- **Improvising with one lane**. Most people who grew up in rural America are familiar with one lane roads and bridges. They're a low-cost, old-fashioned, practical, and proven solution for tight places or dealing with obstacles.

Katie Kam and I are working on a project right now where engineering around existing drainage pipes is prohibitively expensive. So we are working around them by narrowing the protected path to one lane where pipes are in the way. I value the sense of community when you wave through other drivers and they're courteous in return.

You might find companies will help with these expenses. For example, an employer or retail center enthused about Tiny Transit might help defray the cost of constructing bollards or other enhancements to LEAN safety at its entrance. A downtown hotel might help defray the cost of more attractive retractable bollards at its entrance. You'll find an array of barriers on the Tiny Transit Strategies website, from cheap, utilitarian barriers to interesting bollards that could pass for public art.

I'll close with this thought. You can't be everywhere at once. I've given you way too much to do.

So, reach out to people who can help you spot all of these opportunities – people who are seeing new building plans and developments come down the pike. People who will be frank with you. Invite them to be your sounding boards. There's no commitment other than to give you feedback, pass along ideas, and perhaps introduce you to someone who can help.

"People are always coming to us with crazy transportation ideas that cost insane amounts of money and won't work," says Travis County Commissioner Gerald Daugherty. "This idea isn't any crazier, it doesn't cost much, and it might actually work." From Gerald, that is high praise. He's known as a fiscal hawk on transportation.

"Building with crumbs" is actually an advantage. Figuring out low cost approaches sparks enthusiasm in your community. People will appreciate that you are as just careful with taxpayer money as they are with their own money. People will want to pitch in.

This is how you build a groundswell.

The Fax Effect

We are beset by chicken-and-egg questions. What if we build LEAN networks and no one uses it? Or what if the disparate pilot projects never become connected so that instead of a network, we have isolated nodes?

Kevin Kelly, a New Economy guru, identified a phenomenon he calls the "fax effect." Malcolm Gladwell explains.

> The first fax machine ever made was the result of millions of dollars of research and development and cost about $1,000... But it was worth nothing because there was no other fax machine for it to communicate with. The second fax machine made the first fax more valuable, and the third fax made the first two more valuable, and so on. "Because fax machines are linked into a network, each additional fax machine that is shipped increases the value of all the fax machines operating before it," Kelley writes.
>
> **–Malcolm Gladwell**
> *The Tipping Point: How Little Things*
> *Can Make a Big Difference,*
> Afterward, 2002

The "fax effect" also figures in to LEAN networks. The first pilot project demonstrates LEAN Lanes. People quickly see the common sense of it. They want to try it out, which is now easier with shared scooters and bicycles. They don't need a financial analysis to see that it didn't cost much. But beyond the pilot's origin and destination, where can they go?

There will be a next project. Once a second project is connected, the first one becomes more valuable. A third project makes the first two more valuable. And so on.

Aerial and on the Ground

In Austin, we identified fifteen possibilities in locations all over the city. To develop this list, Katie and I spent an intense three days in a sort of mini-charrette. We considered a number of criteria: suggestions we'd received, origins and destinations, traffic congestion, safety of bicyclists or pedestrians, width of shoulders, potential for off-road paths, presence of bike lanes, populations densities, off-road path potential, road capacity, road design, City Council district maps to ensure geographic distribution, and more. We used Google maps and our own tribal knowledge. Feel free to contact me for a link to our album.

The maps made great props for discussing LEAN Lanes with people. Nearly everyone we met with has recognized locations they know well, and they can begin to imagine how it could work.

You can also see things aerially that you can't see at ground level. I found a sixty-foot-wide piece of property owned by a major employer that could open up an entire neighborhood to connect with a possible LEAN bridge across a highway. I'd driven right past the narrow opening without realizing what I'd seen.

Another value in sharing the concept in this way is that the unique profile of each location practically jumps off the page. The environs of a college or corporate campus are different from a downtown public transit hub, and both are different from a community of an upscale neighborhood separated from groceries by a deadly, high-speed road. People understand that the optimal solutions cannot be the same – only the guidelines are the same.

But here's what happens next. The burst of exponential value happens when these site-specific projects are connected to become a network. Synapses connect routes and neighborhoods. That is the fax effect. That is when twenty percent or more of conventional vehicle trips can shift to low emission modes.

Changing the Geometry of Transportation

You might wonder, if protected networks for low speed modes are so great, why is this the first time you've heard about them? If they can improve mobility for the cost of crumbs, why aren't cities already embracing them? If all of this is so obvious, why do *you* need to be so persistent in introducing them in your cities?

First, this is a paradigm change, and paradigms don't change easily. Inertia is a powerful force. Consider how difficult it was for Malcolm McLean, who invented the shipping container. In the 1930s, he was a young trucker trying to get home for Thanksgiving who found himself stuck for nine hours at a shipping dock waiting for the trucks ahead of him to get their cargo unloaded and stuffed piece by piece onto the waiting ships.

> *McLean was no expert in boats. He was a trucker. But he thought this was crazy. He asked the foreman, why not just take the whole trailer off the back of my truck and put it on board the ship rather than unloading and reloading everything?... But everyone laughed at him... It took Malcolm McLean twenty more years to convince everyone else that his idea was right. He eventually had to buy his own boat company, invent a whole new system of cranes... When he was done though,*

instead of taking a week to load, the world's first container ship required only eight hours.
– Charles Duhigg, Arron Byrd, and Samantha Stark
"The Power of Outsiders," *The New York Times*
Video, December 2016

I've heard transportation professionals say, "You can't change the geometry of transportation." But you *can*. Really. Malcolm McLean did.

On the *New York Times* website, you'll find the video featuring Charles Duhigg, a Pulitzer Prize-winning columnist. It reminds us that there may be solutions right in front of us that trained professionals cannot see and might even laugh at.

"Tiny Transit" is a purposefully light and welcoming name. Paradigm changes are difficult enough for people without them also being deadly serious.

Paradigm changes are also easer when you're not alone, when you're part of a movement. That's why we emphasize *community*. Where there is community, there can be a grassroots movement. You can be on the cutting edge together without being on the bleeding edge alone.

The second reason you might not have encountered this concept has to do with the question, *"What problem are we solving for?"* It's a classic question for entrepreneurs. There are Silicon Valley companies clamoring to get into the "transportation space." The problem they are solving for is their own, to create the next "unicorn" – a company that reaches $1 billion valuation. Last year twenty-three U.S. companies became unicorns. Two were in the transportation sector.

I wish all of the companies well because they are coming up with wonderful innovations that can improve our lives. They aren't sitting back. They're investing themselves and taking risks. I'm crossing my fingers especially for two of them that I've personally talked with. I'm in awe of their innovativeness and the good they can do for our world. They are the best of what Whole Foods founder John Mackey calls "conscious capitalism."

But what excites me is collaborating with others to solve some of the world's most intractable social, health, and environmental problems – foremost among them, carbon emissions and climate change.

When our children ask us what we did about the greatest threat of our lifetime, I want for all of us to have a good answer.

LEAN networks change the geometry of transportation

Chapter 7:

Your Big Tent

Reducing carbon emissions or even improving mobility won't be the top priorities for everyone. And they don't need to be. Educate people of course, but also listen for what really matters to them. It's not a competition of causes or a debate about which needs are more urgent. It's about building a grassroots movement that dramatically and rapidly reduces your community's carbon emissions – and more. I call this collection of people with diverse interests "your big tent."

Don't 'Stay in Your Lane'

Your big tent will include people with priorities as diverse as their personalities. Consider these:

- Air quality, of course
- Climate change, of course
- Low cost mobility for all
- Sustainability

- Reducing the impact of transportation's supersized footprint on habitats
- Making active modes like walking and bicycling safer
- Applying big data to make our roads safer and more efficient
- Reducing traffic fatalities and life-altering injuries
- Reducing tension and stress in families
- Allowing people to reduce their carbon footprints
- Improving public health
- Ending addiction to opioids
- Clean air to breathe
- Economic opportunity and inclusion for those left behind in our economy
- Providing dignified ways for people to earn a living
- Reducing the cost of living for struggling families
- Access to education and college classes
- Independent mobility for wheelchair users
- Relieving loneliness
- Economic resilience
- Rebuilding disaster-stricken communities in smarter, greener ways
- Reducing dependence on foreign oil
- Facing climate change as a threat to national security
- Creating more joy and less loneliness in the world

I've heard people say that they need to "stay in their lane" as if it's a good thing, but these are extraordinary times with enormously complex problems. To stay in your lane will almost surely limit your effectiveness and hamper your success.

The most serious limitation is for transportation to be in a lane all its own. Because you need demonstrations for the public, pilot projects, and a groundswell of support, I suggest that you seek out cities with transportation officials who are open to this paradigm change, who share an inclusive philosophy, and who will do the work to create bonds of trust and mutual support.

One way is to reach out to people on different paths and discover the ways you can intersect and support one another's mandates. If you are not naturally outgoing, if you let hurdles discourage you, it's just going to be more difficult, but you can still do it.

Start an ongoing list of people and organizations working in each of the spheres that touch the above list. One by one, get to know them, and through that process, find those who want to build working relationships. Go to events outside of your professional field. Penetrate other groups. Recharge your curiosity. You only need a few allies to elevate what is a nonattainment problem into a broader tent where social equity, cost of living, public safety, public health, habitat protection, economic development, economic resilience, sustainability, and a host of convergent issues combine to build a stronger whole.

Change of this magnitude requires people with "a bedrock belief that change is possible, that people can radically transform their behavior or beliefs in the face of the right kind of impetus," as Malcolm Gladwell writes.

If there are communities that have survived disasters, whether natural or economic, you will probably find that they are very receptive. You are providing them with another tool for recovery.

Listening is the key to building alliances. *Listening* is the key to achieving dramatic reductions in carbon emissions quickly. *Listening* to what matters to people will help you create a grassroots movement that will be unstoppable.

You Need a Bigger Tent

"You're going to need a bigger boat."

The sheriff in *Jaws* utters this line moments after coming face-to-face with the great white shark with its jaws wide open. In a split second, the sheriff's paradigm shatters.

Which is to say, you are going to need a bigger tent. And everyone will have their own reason for being a part of this.

If you host an event to inform and educate people on LEAN networks, NEVs, but you call it a meeting on "nonattainment," you are failing to take advantage of the power of your big tent. People are looking for what *they* are interested in, for what will help them with *their* mandates or passions. To cast it all in the light of your organization's goals is to miss the potential for LEAN networks to solve a broad array of issues tugging at your community and tearing it apart.

Think about *anything* you do and how motivations may differ. Say you are coaching your kid's soccer team. You are there, let's say, because you want to bond with your child. But think of all of the reasons a soccer coach might have.

- To help his child's team perform well so that she's part of a winning team.
- To develop her athletic abilities so she has a chance at a scholarship to college.
- To keep her out of trouble by staying busy with healthy outdoor activities.
- To get to know other parents and make friends.
- To demonstrate team-building to his daughter.
- To teach sportsmanship.

You see the picture. The list goes on. Any one of these could be a parent's primary reason.

Gaining support for LEAN networks involves the same principle. You might think everyone would be much like you, with the same drives and motivations. But in fact, there will be at least a dozen reasons that are as important to people as reducing carbon emissions is important to your work.

This is marvelous, really. It suggests that there is a web of reasons, of slender threads that woven together are stronger than any thread alone. If you were a spider, would you want to spin a taut web that entangles prey? Or have a single thread dangling and hope it works? I hope you go for the web. This is one question that *does* have a right answer.

You don't have to convince everyone else of your mandate and its extreme, even dire importance. Honestly, I wouldn't try. It's more important to listen carefully to theirs.

Let me give you examples of people who are sounding boards for LEAN networks, each with different reasons.

Gary Farmer is an economic development-minded business leader and former "Austinite of the Year." He has helped us because he sees Austin's booming economy has generated a traffic congestion crisis, and it's not possible to build enough roads to solve it all.

U.S. Congressman Michael McCaul chaired the U.S. House of Representatives Committee on Homeland Security when we met. He understands the Pentagon's evaluation that climate change is a looming threat to national security. He and his wife have five children including triplets. Helping with this initiative would make him just about the coolest dad ever.

Catherine Crago Blanton works for the Housing Authority of Austin. Her charter is to develop resources for public housing residents to improve their lives. Catherine is a sounding board because she knows firsthand the financial toll our current transportation system imposes on people who are struggling to gain a foothold.

Joan Marshall, former executive director of Travis Audubon, is a sounding board especially because LEAN networks can reduce carbon emissions as well as reduce the impact of the transportation infrastructure on habitats.

Marion Douglas, a nurse, helps because LEAN Lanes could mean that "my children can never take away the keys to my self-driving tiny car!"

A number of sounding boards now share their perspectives, send us items that catch their interest, ask great questions, and help us strategize. Dozens of their names appear in my acknowledgements. They have helped us develop the language and talking points that you see throughout this book.

The key to our rapport is that we listen to what is important to people and find ways to connect.

Perhaps you know the "Medici effect." It says that powerful innovation happens at the *intersection* of different disciplines. When you begin to explore the intersection of different perspectives, you can generate extraordinary solutions to problems.

This is the power of the big tent.

When Pigs Fly

Peachtree City's multi-use paths began organically when teenagers started driving golf carts to high school, and the city decided to embrace the movement rather than ban it.

Kids are the original creative problem solvers.

Scientist Cristal Glangchai founded VentureLab to teach children entrepreneurial skills as a way to spark interest in science, technology, engineering, and math (STEM). One of her exercises is to ask classes to think of ways that pigs could fly. In a few minutes, her classes can think of more than forty ways. Parents' groups cannot keep up.

Imagine the energy, creativity, and tribal knowledge that kids could bring to the big tent. They have the most at stake in our messed-up world. They will not sit quietly while adults kick the carbon emissions can down the road. They belong in the big tent.

Into the Mouth of the Lion

Conventional wisdom says to build your grassroots movement by finding people of like mind. That's fine, but it's not enough.

You need to find people whom you might expect to *oppose* LEAN networks. You need to call them, meet them, court them, recruit them, or at least gain a mutual respect that neutralizes them.

It's not that hard. In fact, it's one of my favorite things to do. You can meet fascinating, passionate people in this way. I never cease to be surprised at how willing presumed opponents are to meet, work with, and help us.

I call it walking into the mouth of the lion.

These lions won't bite. Far from it. Think of them instead as the lion in *Aesop's Fables*. As a young boy, Androcles helped a lion by removing a thorn from its paw. Years later, having gotten in trouble with the Romans, Androcles is thrown into the coliseum before a bloodthirsty crowd, and a lion is released to devour him. That was considered entertainment. But as the lion gets close, he realizes that the young man is the boy who helped him years earlier. The lion doesn't attack. Instead, he rushes up to lick Androcles. It is an event so remarkable that the Roman emperor immediately releases Androcles.

I can't promise that your presumed opponents will be licking you. But chances are, they will respond to you well if you rid your mind of your preconceptions about what they care about.

Think a moment. Who do you imagine might oppose LEAN networks? Who is skeptical or cynical about local government? Ask people. "Who should I *not* waste time talking to?" Some might be elected officials. They might be the authors of guest editorials in local newspapers, often on the "against" side of an issue.

Make a list of presumed opponents. Then call the most influential, visible, connected, and passionate ones on that list. Reach out to them. Right away you will find them interested, at least in a traffic-choked city like Austin. I believe everyone has some personal expertise in mobility since we spend so much of our time stuck in traffic.

But before you meet, do a heart-check. You must reach out with warmth in your heart. Your goal is *not* to convince or convert them. It is to *listen* to them, learn what they want and what they fear, and establish rapport.

LEAN requires public funding, like roads. I've found that taxpayer advocates are willing to weigh that investment because this is a common-sense mobility alternative with an infrastructure that is a fraction the cost of behemoth transportation projects. They like that you can create flexible LEAN Lanes practically overnight using low cost, prefab barriers, and move them as you learn how people use them. They like our "let's put it out there" approach to testing market demand. I'm right there with them. The only people who are surprised that LEAN networks appeal to taxpayer advocates are the people who've never much spent time with them. That they tend to like the LEAN concept says a lot.

The diversity of interests in your big tent creates its own energy and momentum.

Building consent takes time and patience.

Time Is Precious

The time has come for a mobility alternative that that can help cities everywhere improve air quality while also improving mobility. This positive solution seems more obvious with every passing day. We have a recipe for success to offer you. It will enable you to build your big tent and actually relax because the work is not entirely on your shoulders.

You are amazingly lucky. Your work can improve lives and health and maybe save the world. How many people can say that? Never forget that you and the cities you work with have the power to change the course of our world.

Cutting carbon emissions is possible, and the way to accomplish that is to create an overly-large, welcoming tent where people gather to join forces.

Messier but Easier

Creating your big tent is a great advantage for you. It takes the pressure off of you and your member cities to get everyone on board in exactly the same way, exactly on message. That's too simplistic. It's just not possible.

Instead, listen and ensure that other perspectives are integral to the mobility plans you develop.

Your big tent is organic, growing, shifting, and strengthening. It becomes an energetic, complex organism, a coalition with broad support, real muscle, and inseverable bonds.

Your job becomes a bit messier, but it's easier. You aren't the only one pressing to get LEAN networks incorporated in municipal and regional mobility plans. You'll have an entire team, mutually supportive, arms locked. This is what the most effective leaders do, including William McRaven, former leader of U.S. Special Operations Command and former Chancellor of the

University of Texas System, who attributes his achievements to his life philosophies, including this one.

> If you want to change the world, find someone to help you paddle. You cannot paddle alone... Make as many friends as possible and never forget that your success depends on others... Measure a person by the size of their heart.
>
> **– Admiral (Ret.) William McRaven**
> *Make Your Bed: Little Things That Can Change Your Life and Maybe the World*, 2017

With others helping you paddle, with their hearts in the effort, you can successfully implement LEAN networks in a fraction of the time and cost that major transportation project would require. People of all stripes will join your big tent, enlarging it and diversifying it at once.

Chapter 8:

Your Allies in Public Health

Not only will public health officials want to work with you, they will come with a passion. People working in public health are informed, dedicated, and often frustrated because public health problems can be so endemic and solutions so elusive. Virtually everyone in public health dreams of a homerun that can dramatically improve lives and health. That's why they chose their career. They are naturals for your big tent.

There is no silver bullet for public health. But there may be a bronze bullet.

LEAN networks make it possible for your member cities to improve public health by offering a safe, low cost mobility option that can work for anyone, young or old, rich or poor. But people may not be poor for long! This mobility alternative can help lift people out of poverty.

I recently noticed an interesting aqua Honda seated scooter, Vespa-style, moving along one of Austin's busiest streets. I noticed how careful the driver was. As I sometimes do when notice interesting kinds of vehicles, I followed him. When he pulled into a thrift store, I stopped, jumped out of my car, and caught him before he went inside.

His name was Collin. I was surprised to learn that he had been living on the streets for a long time. It was a hard life. He'd gotten by with help from the wonderful volunteers and donors with Mobile Loaves & Fishes (MLF). He had been considering moving to Community First! Village, the community created especially for the formerly homeless and disabled. But what he'd really wanted was a job making enough money to support himself and get off the streets.

Two months earlier, he leased the Honda scooter for $138/month from Ramble Scooters, including insurance and maintenance. His only other cost? Gas, at 100+ miles per gallon.

His initiative atop a tiny Honda scooter enabled miraculous changes in his life. Now he works for Favor. He was making deliveries when I met him. Now he has a place to live. The income enables him to provide for his basic needs. It's a financial struggle. But he is healthy, happy, and thinking about his future. He is successful in every way that matters. He is a great guy who now is awarded the respect we all deserve. This is the transformative power of low cost, low speed mobility.

Now multiply this one story by the millions of lives that can be improved with low cost, low speed mobility, and you see why public health will be your ally.

Intersections of LEAN Networks and Public Health

You will need to do a bit of research to identify and learn about public health agencies in your COG, MPO, or region. You and your coworkers may already know the public health agencies, their directors, medical directors, key staff, and top priorities. If so, congratulations! You have a head start. If not, this chapter provides essential information for you.

If you're a list person like me, create a list of each agency with what you can discern from its website.

- What community health problems are their top priorities?
- Does the agency view air pollution as a health concern? Agencies in nonattainment cities will probably see it as very serious. What about the others?
- What are the leading causes of death where you live?
- Where do Motor Vehicle Deaths (MVDs) fit in?

MVD is a term used by the Centers for Disease Control and Prevention (CDC) for what most of us call traffic fatalities. At times in this chapter, I use MVD because it's the language of public health.

By researching your region's public health agencies, you can learn how they think about public health problems that LEAN networks help to solve. Chances are you will find at least three major intersections with LEAN networks:

1. **Air quality and the health problems it can aggravate**, including asthma, emphysema, chronic respiratory diseases, and lung cancer. Noxious emissions can also damage the heart, brain, nerves, liver, and kidneys.
2. **Motor Vehicle Deaths (MVDs),** especially injury deaths for young people. This is where you can also consider life-altering injuries.
3. **Social determinants of health,** the term used by public health professionals to refer to the myriad of conditions in people's lives that shape their health. Senseless injuries, depression, suicide, stress-related illnesses,

opioid addiction, chronic diseases, constant pain – it's all overwhelming, interconnected, and beyond the ability of government to solve – or is it?

Let's look at each of these intersections. I would like for you to think about the effects of mobility on people's lives and health so that you will see why public health officials can become full-fledged partners in your big tent.

Rest assured that you do not need to become an expert in public health. Far from it. You simply need to think broadly enough to include, communicate, and – best of all – collaborate with public health experts, officials, and foundations in your cities.

Health Impact of Poor Air Quality

If you work at a COG or MPO, this is a slam-dunk for you. Your region's air contains some level of noxious emissions, whether a little or a lot. LEAN networks will help move the needle on noxious emissions whether your counties are in nonattainment or struggling to stay safely out of range. As our sounding board Gerald Daugherty says, "Who wants dirty air?!"

Life-Altering Injuries

Second, I discussed Vision Zero and *safety, safety, safety* as an imperative in Chapter 5. Now I will shine a light on life-altering injuries and the long-term damage to people's lives that LEAN networks can help prevent.

Cities and counties are great at tracking traffic fatality numbers, but I have yet to run across a city or county that measures and tracks longitudinally the secondary health effects of traffic collisions.

So, during the nine years I spent studying traffic fatalities, I came up with three rules of thumb. Every city's numbers will be different, but you can't go wrong by using these simple formulas. Public health officials will recognize that these are conservative estimates, so you don't need to waste anyone's time justifying them.

1. **MVD times 4.** For every traffic death, there are four life-altering injuries.
2. **MVD times 40.** For every traffic death, there are forty injuries of all kinds, from minor to severe. For example, I visited a long-term rehab unit in Austin where traffic injuries accounted for fully half of the dozens of patients.
3. **MVD times 2.** For every traffic death, there are at least two personal bankruptcies or catastrophic losses. These may be triggered by the loss of a wage earner, loss of a home to foreclosure, medical costs, suicide, or care-giving. If you know that a city has one hundred traffic deaths annually, you can reasonably estimate that there are at least 200 cascading losses that slam through thousands of lives and families.

When you think of it this way, the magnitude is staggering.

We don't ordinarily encounter people suffering from life-altering injuries unless we are among their immediate families or close friends. We never again see those who died. Survivors of traffic collisions live with pain, disfiguring burns, disabilities, medical issues, depression, and isolation. A family can be consumed by caregiving and bills, barely scraping by if their breadwinner can't work. No safety net is enough to save them from continual hardships. Survivors and their families rarely come to public hearings or neighborhood meetings. They may have dropped out of the workforce because of disabling injuries or because their disabled family member needs care 24/7. They may have lost their homes and roots. My hero Alan Graham, who spent nights on the street getting to know the homeless in Austin, tells us that the main cause of homelessness is catastrophic loss of family.

When we encounter these survivors, we empathize with them. But sadly, they are mostly invisible to us. I have a friend who cannot even think about such horrific losses, and I understand and respect that. Thinking about them reminds us that we are all just one traffic collision away from falling into the forgotten abyss.

I try to incorporate the pain and loss of these families into my thinking because it energizes my work. I try to voice what they would say if they could be heard. I am outraged that they are society's cast-offs. They cross my mind at some point every day. They are top-of-mind when I'm working on this project. Whenever I feel that this passion of mine is just too much of an uphill fight, that we are too wedded to our conventional vehicles and roads to change until it's too late, I stop and remember that I am speaking for them, saying what they would say if they could speak. Our work is so much easier than what they and their families must live with. I know of a young man paralyzed from the waist down from a car crash when he was five years old. I know of a woman now in her twenties who will never be self-sufficient because her drunk mother ran into a truck when she was a toddler. I know of a middle-aged man who was paralyzed decades ago when he was struck by a car on a high-speed chase. Now and then you will see an obituary about a person who eventually died from crash injuries suffered years before. That person existed but is seldom counted. That person doesn't show up in the Motor Vehicle Death data. He or she doesn't really show up anywhere. No one died at the time, so there is no vehicular homicide investigation.

Public health officials know these survivors exist and that the precipitating event is often uncounted. For example, how do you count an opioid death with an addiction that began with prescription painkillers following a debilitating car crash? No one can undo their injuries, eliminate their pain, or put their spine back together. The least we could do is count them.

Maybe you know the old joke about a night patrolman finding a man searching the ground under a street lamp. The policeman asks, "What are you looking for?"

"I lost my watch," says the man.

"Where did you lose it?" the patrolman says.

"Over there." The man points a half-block down the street to an area that is pitch dark.

"If you lost it over there, then why are you searching here?"

"The light's better here," the man says.

In the same way, we measure fatalities because they are easy to track and easy to see under the light, whereas we don't know nearly as much about life-altering injuries. These are as important as fatalities but more complicated. It is not clear at the moment of impact what the ultimate health effects will be. The injuries might heal, or they might last a lifetime, derailing careers, marriages, and families.

Barring a miracle, a child's brain injury will follow him for the rest of his life. Life goes on – for everyone else in his class. This boy grows up little noticed, falling behind, barely remembered. There is no medical breakthrough to right that catastrophic wrong. Life goes on but at an enormous, unrecorded human cost.

Life-altering injuries constitute a public health crisis.

We have the power to solve this. We know what it takes. We know that speed maims and kills.

Social Determinants of Health

Third, I will devote the rest of this chapter to the social determinants of health. This is where I will weave all of these public health problems together. You will see why public health allies are crucial in your big tent and how you can win them.

"Social determinants of health" sounds like bureaucratic jargon, doesn't it? I'm afraid that it's all we've got to describe the myriad of factors that affect people's health and lives. It is the language of public health.

Without low cost mobility, life is harder. Getting to jobs, job interviews, classes, groceries, medical clinics, family, and friends is hard without a car in our auto-centric cities. In many cities, from Atlanta to Austin to Los Angeles, a person without a car can be *persona non grata*.

You know this. You see this.

Think for a moment of all the issues that are plaguing our cities. This is your quick immersion into the social determinants of health.

- Poverty – Even in cities that are prospering, poor families live just across a freeway or crammed invisibly into a small house.
- Low wages – Inadequate to support an individual, much less a family.
- Hunger/food insecurity – Many children in our cities go to bed hungry.
- Addiction – Across the country, opioid deaths annually now exceed MVDs.
- Chronic illnesses and pain – Going untreated or stretching medicines due to cost and lack of access to care.
- Affordability – With its harshest effects on low- and middle-income families.

- Lack of affordable housing – Public housing is scarce, and market-rate housing is too expensive.
- Homelessness – Every city has homeless people camouflaged against buildings and under bridges.
- Unemployment – People lacking the education, skills, or means to get a well-paying, stable job.
- Isolation and loneliness – Especially among, but not limited to, the elderly.
- Undereducated – Many people lack job skills, and better jobs aren't open to them.
- Social inequities and access – Many people are left behind in our economies.
- Mental health – Depression and mental illness are often untreated.
- Unnecessary deaths of all kinds – Whether illness, self-inflicted, violence, or accidents.

Are you struck by how interconnected and overwhelming these issues are? Me too. What's encouraging is that a low emission mobility alternative can be a game-changer. LEAN networks can:

1. **Lower the cost of transportation,** allowing low income people to participate more fully in their city's economy. LEAN networks support low-cost modes for people to get to transit, jobs, education, groceries, health care, and family members. Remember, NEVs typically are one-quarter the cost of owning, operating, and maintaining a conventional vehicle. Other low emission modes cost even less, some far less.

2. **Connect with public transit**, solving the first/last mile shortfall with a flexible, low-emission solution.

3. **Relieve financial stress,** reducing depression, easing stressed family relationships, dialing back domestic conflicts, and ending isolation and despair.

4. **Provide a low-cost platform for new technologies,** making them available to everyone, not just luxury car buyers. Every life deserves to be safe from human error and the deadly epidemic of distracted driving.

5. **Revolutionize access for people with disabilities,** offering independent mobility for wheelchair users. Attorney and entrepreneur Stacy Zoern, who has never walked, developed an NEV that a wheelchair user can drive with their wheelchair as the driver's seat.

6. **Reduce transportation's impact on the environment,** shrinking transportation's footprint, allowing denser development, and preserving green space. LEAN networks also reduce highway noise and noxious fumes that waft over nearby neighborhoods.

7. **Empower small-scale entrepreneurs,** making it possible to start or grow a small business like food vending, landscaping, handyman services, or pedicab operator. I recently noticed a fully outfitted preowned Polaris GEM ice cream truck with less than 1,000 miles on Craigslist for $15,000. Recalling my happiest job ever, serving ice cream at Baskin Robbins, I was tempted to buy it before it disappeared. That purchase can become the basis for a business or second job.

8. **Strengthen economic resilience** so that cities are less vulnerable to the next recession, spike in unemployment, or hike in oil prices. The stock market can be volatile, with dramatic swings. What happens when the boom ends? Families are more resilient when they're not handcuffed to

a $400+/month car payment and costly maintenance and gas just to get to work. What happens if they're laid off or furloughed? LEAN networks enable families to live on less and pull through tough economic times.

Polaris GEM six passenger shuttle

Community First! Village

I mentioned that Katie Kam and I are developing a mobility plan with LEAN Lanes for Community First! Village, a fifty-plus acre community for the formerly homeless and disabled, located on the eastern outskirts of Austin. We've been retained by Mobile Loaves & Fishes, the sponsoring organization, led by Alan Graham, the practical genius who founded it all.

You might think, hey, wait a minute. Do we really think that a mobility choice for the formerly homeless would appeal to middle- and upper-income people who have more money to spend and are conscious of social status?

Yes, that is exactly what's happening. LEAN networks are the great equalizer, just as a Coca-Cola is exactly the same whether a billionaire is drinking it from a crystal goblet or a construction worker is drinking it straight out of a can.

There is no stigma. Cool is cool.

The broad appeal of LEAN networks has enormous implications for public health. It lifts the pressure to spend money on vehicles that people can't afford. Necessities are redefined. A struggling family of four with two cars might be able to replace one of them with an NEV or pedal-assist bicycle. Money that might have been spent on gas, maintenance, and auto insurance could be better spent on kids' athletics, braces, or college tuition. Reducing the cost of mobility makes it possible to smile on a tight budget. Cutting the cost of transportation in half or more is like getting a raise. It's a mood elevator without a prescription. You know Pharrell Williams' song, "Happy"? This is happy.

Mobile Loaves & Fishes is developing a LEAN mobility plan for Community First! Village

You need public health allies, and guess what? They need you, too. LEAN networks can help them accomplish what seems impossible.

Chapter 9:

Your Allies in Social Equity

LEAN networks are a game changer for social equity. If there were no other reason to create safe passage for low cost, low emission modes, social equity is reason enough.

We are struggling in the U.S. to create opportunities for *everyone* to participate in the economy, for *everyone* to be able to afford to live in safe, adequate housing. We want children to have breakfast and families to be able to get to grocery stores and buy enough fresh food to feed themselves. We are a country of plenty, yet we have had to coin terms like "food desert" and "food insecurity" because so many Americans live their lives in chronic hunger and anxiety that they simply cannot afford to put food on the table. Even when they have enough money to buy food, how will they get to a grocery store that is far away and then carry back the sacks?

How can parents without transportation get to classes so they can qualify for better jobs? In many U.S. cities, a family without a car is simply a discard, a casualty of our economy. Say you want to help one of these families. Where would you even start? Social service agencies work with struggling families every day to help them through a myriad of obstacles. For many families, the cost of transportation tops the list.

- In Austin, a city of plenty at the top, many low-income households must spend well in excess of twenty percent of their incomes on transportation. They have no choice. Their employment depends on it.

- You might think that the greatest barrier to moving out of public housing is finding affordable market rate housing. But in fact, the cost of transportation is the greatest barrier for many such families.

What does this mean to you, trying to cut your city's carbon emissions? *Everything.*

You need alliances to build a groundswell in your city. If yours is a mid-sized or large city, chances are that it has dozens of agencies and foundations dedicated to helping families in need. Groups dedicated to bridging the income gap, helping families get on their feet, working to improve the social determinants of health.

You may already know some of your city's leaders in social equity. That's a head start. But you can also start cold. Your assignment is simply to find a few leaders in your city's social equity/social service community who will be inspired by your inclusive vision of your city's future. These are your magic words:

Our vision is a world in which all people can get everywhere they need to go within their communities using this low cost, low emission Vision Zero mobility option. We envision a protected infrastructure, free like sidewalks, that becomes a city-wide platform for low cost autonomous technology innovation.

Think about these few words. They express the radical vision that poor and low-income families can afford mobility, just like everyone else, that we will create community-level mobility options that work for everyone, regardless of income, and that the technologies it spawns can benefit everyone.

Your allies can participate in this vision. No one knows the needs of the bottom rung of the socioeconomic ladder better than the people who work to help disadvantaged people and families every day. Imagine how much more inclusive and inspiring your vision can become when they bring their experience, wisdom, and passion to the mix.

Early in my journey, I visited with the CEO of the United Way agency in Austin. As we met and I described this concept, I noticed that tears came to his eyes because this was a dream come true for the families who are most challenging for its agencies to serve. These are deserving families left so far behind in our city's economy that it's nearly impossible to get a foothold and climb out of the deep, disconnected hole they are in.

I met recently with the CEO of a $200 million community foundation, a collection of foundations and funds with a broad spectrum of interests – children, health care, addiction, the environment, access for disabled, and more. A friend, also at our meeting, asked, "What makes a foundation decide to give money to a cause?"

"It helps if you can make them cry," he said.

In our hearts, we know this, don't we? It is why a single tear in a Keep America Beautiful television commercial in the 1970s still causes our own eyes to water forty years later.

You have something amazing to offer people who are working in the trenches to help their clients who are at the bottom of our society. These dedicated social service providers can help you make this new mobility alternative even kinder, gentler, and more inclusive.

Kinder, Gentler

Striving to be "kinder and gentler" is embedded our American character. Wise, long-range thinkers understand that it is in our national interest to lift those who are struggling on the lower end of our society. Our country, like a chain, is only as strong as its weakest link.

So how do we assure that this new approach to mobility is implemented in a "kinder, gentler" way? How do we ensure that LEAN networks are low cost and inclusive, truly helping people connect with jobs, education, and economic opportunities?

We accomplish this by establishing unified standards. I developed these Kinder, Gentler Mobility Rules as a starting point. The objective is to ease the stresses and pressures of life, especially for those most in need of low-cost mobility. Here are the basic rules.

- Never allow motor vehicles that can exceed twenty-five mph to use LEAN networks.
- Never impose a toll on LEAN networks.
- Never exploit people under the guise of acting "like a business" to generate revenue for any entity, public or private.
- Never prohibit people from using LEAN networks because of a past due payment, fine, or ticket.
- Never institute fees, tickets, or fines that escalate at usurious rates over time.

- Never turn over collections to a third party with an incentive to collect money rather than help people work out matters without financial stress.
- Never tow or boot low speed vehicles unless they are inoperable or blocking traffic, in which case retrieval must be at minimum cost and at a convenient location along LEAN network corridors.
- Always allow people the option to work off any fines or penalties with community service.

Social equity is a fancy word for ensuring that what we do helps to bridge, not widen, the gap between the haves and have-nots. Kinder, gentler mobility rules support social equity.

Network design must be integral to the neighborhoods and communities they serve. LEAN Lanes cannot be designed on a CAD system by traffic engineers in third floor offices located seven miles away without up close, extended on-the-ground observations for the stretches of routes they are designing.

LEAN networks require a more granular way of thinking about transportation design. Think of this as creating a grassroots-informed mobility mode. It is about meeting people where they live and gaining their insights into how to best design flexible protected networks, and then revisiting them to see how it's going and fixing what needs fixing.

It is about listening to parents as they show us how they get their children to school and what worries them. The conversation may jog their memories of dangerous incidents or near-misses that they may have witnessed. It is about working with retail stores in strip malls and restaurant managers with drive-through windows to understand ingress and egress from their customers' perspectives. This is especially important in low income areas where city codes may not be as well enforced and where there is a lot of foot traffic with risks to pedestrians. It is about understanding where and why people jaywalk (or now "jay-scooter") or ignore laws while driving or bicycling. You want observations like this to help create a true Vision Zero mobility alternative that works even in the poorest and most stressed neighborhoods.

Collaborating with Nonprofits

So, you now know that you have an opportunity to build relationships with social service agencies and community leaders committed to social equity. These must be relationships of mutual trust. What I am about to tell you is essential to that trust.

You and your grassroots movement cannot step on another cause on the local scene competing for foundation and donor dollars. You cannot create another nonprofit or a competing cause and then expect for social service agencies to be your allies. Money is not your objective. The LEAN network is.

Governments – local, state, and national – pay for our transportation infrastructure. That's the way it should be. If governmental entities tell you that you cannot afford to build the LEAN infrastructure, then you have failed to communicate to them that *they cannot afford not to build it.* It may mean you need to sharpen your city's version of "How LEAN Networks Can Be Built with the Crumbs of Major Transportation Projects." Believe me, I know how hard it is to get someone to suddenly see that there is a dramatic opportunity to offer low cost mobility *to all* when they have spent decades with the present system and have its resolute inequities stuck in their minds.

Infrastructure is not a charity. It is not a cause. It is a mandate from our planet to stop carbon emissions before it is too late. Building that infrastructure, and quickly, is the job of government.

Cities can best help nonprofit organizations by enlisting their involvement to expedite LEAN networks. Cities can ask nonprofit organizations, "How can we help provide access to LEAN networks for the people you represent and serve?" Here are a few directions this conversation could go.

- Nonprofits can identify pilot projects and request that government expedite funding the infrastructure.
- Either nonprofits or cities can propose to your city's transit authority to cordon off a pilot infrastructure route and provide a fleet of shared low speed modes for underserved neighborhoods.
- Cities can open doors with lenders to create a pool of community funds that low income residents can borrow on favorable terms so that they can lease or buy Neighborhood Electric Vehicles or other low speed modes.

As the LEAN infrastructure comes into being, as it begins to exist and serve people, it will create enormous opportunities to do good, and quickly.

In short, for LEAN networks to be implemented equitably, it is essential that you nurture relationships with social equity advocates and leaders who will naturally bring their passion to creating the low cost, low emission, low speed protected infrastructure that serves everyone.

Building the infrastructure is *government's* role.

How the infrastructure improves lives while cutting carbon emissions is *everyone's* role.

Chapter 10:

Big Data, Meet Tiny Transit

We are now in the midst of the next tectonic shift in transportation. We have begun the transition from the internal combustion engine to whatever is next.

No one knows for certain what it will be. American cities were taken by surprise in the spring of 2018 when companies dropped dockless scooters onto the scene like gigantic confetti. Who among us could have predicted how quickly these dockless modes become pervasive? What's next…flying cars? …autonomous roller skates? How can you help your cities steer toward an efficient, low emission future when there are so many unknowns?

One of the most exciting frontiers is big data. Our children are growing up in a world where they take for granted that the television will not only deliver a movie in high definition on demand but also recall precisely where they paused it and ask if they want to resume or start over, yet my son can be struck by a car in a crosswalk on his way to get gelato.

Clearly, we have a lot of work to do. Your challenge is to help your cities transition to a low emission future and to anticipate and seize the opportunities that will come with big data when you don't know exactly what lies ahead.

The Big Bang of Big Data

I confess that I am having a hard time imagining what is next in our explosive, data-rich era. I sincerely want to know. Experts tell me I'm not alone.

We are collectively witnessing how rapidly big data has swept entire industries and become embedded in daily life. Apps are pervasive. Rare is the job that doesn't require digital knowledge. Big Brother has entered our homes and lives with deceptively innocuous names: Alexa, Siri. From global communications to medicine, big data enables miraculous transformation. It can make our trivial tasks easier while collecting massive amounts of data about us.

Two of the most valuable companies in the world – Apple and Google – possess and monetize data beyond comprehension, and both are exploring how to upend transportation with data-rich innovation. And of course, data is Uber's enabler and raison d'être.

Yet for all of that data, roads are still the deadliest place for young people to be. Half as many people are killed on our roads each year as American soldiers in the entire Vietnam War. There is an enormous disparity between the sophistication that data brings to our most mundane tasks, and the "every car for itself" carnage on our roads.

This is where LEAN networks come in. The LEAN infrastructure is an ideal platform for low cost autonomous (self-driving) and connected technologies. You don't need a $40,000 to $80,000 vehicle to benefit from these technologies.

Recently, I learned that the technology exists to keep shared low speed vehicles in fixed, invisible LEAN Lanes. It's sort of like an electric fence for dogs, except that instead of getting shocked, the vehicle power cuts off when a driver tries to steer the vehicle out of an approved path. It is essentially a force-field originally developed for the military. Now, it can be applied to LEAN Lanes to keep drivers from veering off the protected course. It can also prevent the runaway shopping cart phenomenon of shared Tiny Transit modes being scattered and having to be collected by modern day cart rustlers.

Big data can help personalize LEAN networks and Tiny Transit modes. Big data can help LEAN drivers work with the challenges of terrain and rights-of-way. Instead of a one-big-size-fits-all approach, big data can allow LEAN networks to offer the opportunities to improvise, working with what exists to ensure "redundant safety" and keep costs low. Big data is the enabler for the low speed, low emission shared modes activated by apps. The fact is, we are only now beginning to glimpse how big data will change transportation.

The City of Dallas Director of Transportation Michael Rogers foresees a rapidly approaching future.

> Municipal governments need to understand the data that we possess. We give away data to companies that package and sell it, making millions of dollars using the transportation infrastructure built with taxpayer money. Why is it that the government is not in tune with what's going on? The dockable scooter and bicycle companies are not in this for transportation. They're in it for data because that's where the money is. Municipalities need to understand that this data has value. In addition, we need to put in our contracts that

this information cannot be sold or used to violate privacy. The data is less valuable that way, but it still has value. We have a responsibility to ensure privacy.

Imagine the "big bang" of big data that will be generated from LEAN networks as they prove to be an ideal platform for low cost smart technologies.

In Chapter 5, I described how can cities can build LEAN networks with the crumbs of major transportation projects. I now believe it is possible for cities to restrict use of that data to ensure privacy and sell that data to help accelerate its LEAN network.

The City of Dallas is looking to universities as prospective partners on technology opportunities. "We need to start training our young minds to start thinking a bit deeper so that we get our next leaders at the cutting edge, and universities can help us better understand the future," says Rogers.

Los Angeles in 2018 unveiled its Mobility Data Specification (MDS), a tool that allows the city to gather and analyze real-time data from new dockless modes. Los Angeles Department of Transportation (LADOT) Chief Sustainability Offer Marcel Porras says, "We can no longer simply rely on those analog tools we have for so long. We have to create a digital set of management tools" for which MDS is the foundation.

Most small and mid-size cities lack the bandwidth to keep up with what's happening in big data. In response, San Diego Association of Governments (SANDAG) is working to become a microdata clearinghouse and resource for the cities in its region.

It's all happening so fast. If you can't keep up, you're in good company.

What We Now Know

True, we don't know what's next. But we do know a lot.

We know that the current system is not sustainable. Our dependence on the internal combustion engine is too expensive, spewing higher levels of carbon emissions than the air in our cities can handle, with hundreds of thousands of senseless deaths and serious injuries annually, all while failing to serve too many people, burdening others with heavy costs, and endangering our planet.

We know that there will be a transition of some kind. We don't know what form vehicles will take.

We know that auto-centric land use is not sustainable.

We know that low cost, low speed, low emission modes will be part of our future.

We know that allowing these vulnerable modes to come into conflict with conventional vehicles is deadly.

We know that on-demand apps will grow.

We know shared services will grow.

We know that many young people today don't need or want to own a conventional car.

We know we need first/last mile solutions for public transit.

We know companies and university research labs are racing to develop big data and smart technologies.

We actually know quite a lot that we didn't know five years ago, or even a year ago.

Your cities will play a crucial role determine what the next era of transportation will look like. Helping to inform cities' decisions will be data, and lots of it – especially the cumulative trillions of bits of information on choices that people make as we go about our days. Suddenly, there are many more decentralized possibilities – NEVs, pedicabs, scooters, Scoots, Segways, Vespa-like scooters, hoverboards, skateboards, and who knows what we'll see next.

What we do today, this week, this year, matters.

Tribal Knowledge

One of the advantages of LEAN networks is that they can use "tribal knowledge." People who live in neighborhoods – the tribe – often have a better understanding of traffic patterns, dangers, and how people actually get around than transportation professionals with their eyes on their CAD screens.

Last year, I witnessed a hit-and-run. A speeding car clipped a bicyclist, knocking him to the ground, and sped off. I still have the license plate number and the phone numbers for witnesses. But because the bicyclist didn't file a report, there was nothing to attach my report to. I learned that there was no system to receive this information. It's as if it didn't happen. Tribal knowledge was lost.

Here is another example of tribal knowledge, pedestrian style. When I was a student at UT Austin, it seemed that every month there was a new building springing up on campus. Whenever one opened, the sidewalk was invariably in the wrong place. Naturally, we students walked the most efficient straight line to the entrance. Then the University would post a sign that said, 'Keep off the grass.' And, when we ignored that, they'd suspend a chain. Then they'd raise the chain height and make it a double chain – which didn't stop us. I wondered, why didn't they wait until a week after the building opened and build the sidewalk where the worn grass paths were?

Tribal knowledge is important to building the groundswell where you live. It will help make LEAN networks organic, valued, optimally designed, and properly protected. There is a sense of ownership when neighborhoods and communities are involved in siting. People are more likely to consent to compromises and impositions, say, yielding property as right-of-way, or sharing the road, or understanding why it is in a homeowner association's interest to share in the cost of an off-road path. That sense of ownership is an intangible that produces tangible results that can accelerate implementation of LEAN networks.

Chapter 11:

Cities as Leaders

What does all of this have to do with your cities? *Everything.* The transition to low speed modes cannot happen without cities leading the way. LEAN networks must be built quickly, cost-effectively, flexibly, and safely. They are the opposite of interstate highways in that they are small, Vision Zero, low cost, low speed, and low emission. They are local.

To fulfill this role, cities will need multidisciplinary LEAN design teams to fast-track LEAN networks. This cannot be "transportation as usual." We can't afford an expensive, massive infrastructure that is over-scoped and takes years to plan. We have to take *quickfire, calculated shortcuts with safety* as the prime directive.

Design cannot be left solely to transportation planners and engineers. They were not trained to think in the ways that we must now think. It is asking too much of them. There can be no transportation silos. We need advocates for fiscal conservatism and experts on public safety, public health, economic development, sustainability, and smart technology alongside transportation engineers on the infrastructure design team, *not to slow things down but to expedite them* and ensure that the networks are safe, efficient, low cost, common sense, and replicable.

We cannot look to the federal government to provide all of the answers. The groundswell won't happen in Washington D.C. It will begin with cities and communities across the U.S. That is why I wrote this book and why I am placing it in the hands of mayors, activists, COGs and MPOs.

Do you know how your brain and nervous system can speed up in the face of a deadline? How some of your best thinking happens under pressure? When you are truly in a creative problem-solving mode under deadline pressure, have you ever noticed that you surprise yourself with the quality of your work product? That is the urgency I want to inspire in you. It's the "power mode" of Medici thinking. It is your secret sauce. You see the problems in our world, and *you have a breakthrough solution* for carbon emissions that cannot wait for everyone to catch up. Instead, we have to demonstrate and prove this concept.

I recognize that the U.S. will need at least ten more equally effective solutions to turn back carbon emissions. We can't let that shortcoming stall us. Instead, let's implement this one solution in a way that will spark ten others.

LEAN bridge enhances placemaking

Discovering downtown Austin's charm by pedicab

The Great Rewrite

There will be tremendous opportunities and profits for the companies that are on the first waves of "The Great Rewrite" – which is what *Forbes* calls the profound trends that are rewriting the way we live and work.

- Who could have foreseen the iPhone twenty years ago? Wireless networks made smart phones possible. Apple, the great and fearless innovator, is now the most valuable company in the world. Likewise, low emission networks will make possible innovations we cannot imagine.

- The railroads missed an opportunity when their executives failed to understand that they weren't in the *railroad* business; they were in the *transportation* business. In the same way, big oil companies aren't in the oil business; they are in the *energy* business and the *chemical* business.

Automakers aren't in the car business; they are in the *mobility* business.

Michael Rogers, Director of Transportation for the City of Dallas, adds this perspective. "In the 1890s, you saw horses. Over a thirteen-year period, we shifted from horses to automobiles. I think we're at another transitional point in our history where we're about to see new innovations that are going to improve lives tremendously."

The smart money is on the companies that get this and the entrepreneurs who are developing game-changing technologies. Let's encourage them. Consider:

- Our economy *not only survived but grew* from the radical transformation of our telephone industry to the wireless telecommunications industry today.
- Our economy *not only survived but grew* when Steve Jobs and Steve Wozniak invented products that brought computing power directly to the people.

Our economy loses *only when it stands in the way of innovation.* It loses only when we allow entrenched interests to shiv innovative ideas.

Cities that step up to become innovators and early adopters can lead the way by becoming test sites for companies that are building profitable new business models by investing in the infrastructure, products, and technologies that will change the trajectory of our planet, making new industries and new job opportunities possible. The new mobility can be a renaissance for American manufacturing, engineering, and ingenuity.

Land Use Is Destiny

Most U.S. cities today are being designed and governed by land use laws and ordinances that were written by the great-grandparents of today's young adults. At the time these rules were written, Americans were newly in love with the massive cars coming out of post-war Detroit. There was a feeling of endless possibilities for paving the landscape and extending it to the new remote suburbs, where houses were spread out, garages were becoming the face of houses, and people drove pretty much everywhere. Gasoline was cheap. To flash back to that time, think of Mad Men and Don Draper's world. Car was king, men commuted to downtown jobs, littering along highways was the norm, and bigger was better.

Wastelands of empty parking lots across our country

Today we have an entirely different world – and we've squeezed it into the urban form just as I might squeeze my size nine foot into a size eight shoe. It hurts. It's crippling us.

No one knows what to do about this because land use is codified in our city's laws. Colleges and universities teach these laws to engineering, architecture, regional planning, and public policy students, whose careers then require them to operate them within those laws. If the students were taught something different, they might well be unemployable in their chosen fields.

Changing the laws is a monstrous task. A few attempts across the country have fallen flat or struggled to alter the fundamental factors in cities' laws.

Yet if we fail to change the way our cities function, we are doomed to repeat every mistake *ad infinitum* until carbon emissions take down our planet, if the lack of economic resilience and frayed social networks do not take us down first.

When I sit in conference rooms and hearings with people who accept to their core this unlivable model, even as they criticize it, even as a brave few try to change it, I realize how formidable this model is. Attempting to change it is like the mythical Sisyphus, condemned for eternity to push a boulder up a hill, only to have it roll down each time. It's easy to see why people give up, give in, and become invested in the same outdated system that may well have quashed their own "color outside the lines" ideas of their idealistic youth.

It is urgent that we cut carbon emissions, yet we are lemmings heading toward a cliff, ignoring cries to change course.

Nothing will change *unless cities step up to the challenge of leading change*.

This is where COGs and MPOs come in to help accomplish things that cities on their own cannot. Acting collectively, cities can move toward a more resilient, sustainable, clean air future. We are seeing this in the innovator and early adopter regions.

- Coachella Valley Association of Governments (CVAG) accomplished something that only a few visionaries and optimists like executive director Tom Kirk and Healthcare District leaders would have thought possible. In 2018, it opened the first two-mile section of a fifty-mile multiuse low speed, low emission path connecting multiple cities and tribes.
- San Diego Association of Governments (SANDAG) does not own land or build roads. It plays a different sort of role. Working with California State Senator Pat Bates, SANDAG sponsored legislation enabling San Diego County or any city in the county to establish an NEV transportation plan to serve the needs of its residents. The legislation received support from the Sierra Club, the Electric Vehicle Association of San Diego, and the Center for Sustainable Energy.

CV Link is beyond what any individual city could have achieved on its own. SANDAG enlisted its member cities in the campaign for enabling legislation. These projects happened *because these cities were working with their MPOs* to lead the change.

Cities Leading the Paradigm Change

Imagine what is possible if cities collaborate to change the calcified paradigm that stands in the way of a new mobility alternative that cuts carbon emissions. Here are two possible ways this could work. I'm offering them to stimulate conversations.

1. *LEAN Monthly Accelerator.* Organizations of cities and smart city groups could convene representatives for the sole purpose of offering draft legislation and ordinances for

cities that want to develop LEAN networks. I envision this as a multidisciplinary team with the best minds among us, including not only public transit, transportation engineers, and urban planners but also creative thinkers in public safety, public health, energy, and the environment, along with the CDC, foundations, innovators like Better Block, and a few McArthur Fellows for good measure. Working with organizations representing COGs, MPOs; the Urban Land Institute (ULI); U.S. Conference of Mayors, and others, the Accelerator could begin to generate new draft municipal laws to override the wave of 1950s-era auto-centric local laws and development codes that are still on the books today.

We can't wait for the entire code rewrite at once. It would take years, and who knows if the product would be effective or even any good? Thousands of localities would have to debate and adopt it with who knows what results. That would be a decade-long process at best. Instead, I envision new draft language each month until all of the constraints have been addressed. This years-long project begins with a step, then the next step. Each step accelerates the groundswell, learning as we go. Cities could opt in or not.

The *LEAN Mobility Accelerator* could help cities change ordinances that limit innovative modes and entrepreneurial solutions. I understand why dockless companies dropped thousands of scooters in U.S. cities without getting advance approval. They were smart to ask for forgiveness, not permission. Recall Chris Nielsen, Electric Cab North America, and the five difficult years he spent getting Austin's permission to operate what is now a multi-city fleet, the largest of its kind.

Imagine how the LEAN paradigm change could be advanced if U.S. cities had a place of their own, by, for, and about cities, to look to for very specific, proven language and guidance for new ordinances to accelerate LEAN networks and micromobility.

2. ***Vision Zero unified design standards.*** New mobility modes are landing in cities like it's the Wild West, when territories were lawless and fortune-seekers were reckless. Scooters have brought a sense of excitement and possibilities. We want to encourage these new modes – *and* ensure safety. The cautionary tale about cell phone use in Chapter 5 reminds us of what can transpire if we fail to do so.

With cities leading the way, the U.S. Department of Transportation and CDC should *develop uniform LEAN design standards that prioritize safety without choking innovation.* Working with cities that are innovators and early adopters, the federal government could help ensure the integrity of a vision that prioritizes safety, contains costs, applies common sense, and remains free of charge for everyone.

For LEAN networks to proliferate, we need network design standards that exceed the unacceptably low safety standards for bike lanes. The design goal should be Vision Zero while increasing safety for bicyclists and pedestrians as well.

If you are interested in more possibilities for cities collectively to lead the way, stay connected through the Institute for Community MicroMobility.

A Word about Trademarks

Three years ago, I filed a trademark application for Tiny Transit™. It's a lengthy process. I expect to be granted a registered trademark, indicated by the symbol ®, in 2019.

I have also filed a trademark application for LEAN Lanes™.

I filed for both as "credentialing" trademarks. They allow me to set standards for their use. For example, to represent a vehicle as Tiny Transit™, it cannot have a maximum speed greater than 25 mph, it cannot be a golf cart, and it must be low emission.

Likewise, for a path that's part of a LEAN network to have a LEAN Lanes™ sign, that section of the path must be protected from conventional vehicles, it must be free for users, and it must comply with the Kinder, Gentler Mobility Rules outlined in Chapter 9. I expect to receive the LEAN Lanes™ registered trademark in 2020.

The reason I filed for these trademarks is to protect the integrity of these terms. Without credentials, anything could call itself Tiny Transit™ or LEAN Lanes™. They could lose all meaning. We have already seen the term "microtransit" morph to include conventional vehicles like Uber, Lyft, and vans used as circulators.

Someone sent me a link to an article titled, "The Story of Microtransit is Consistent, Dismal Failure." What…? But this was about the failure of app-driven large vans and minibuses, not the kinds of vehicles that meet the credentialing requirements of Tiny Transit™.

Even as early as 2015, I saw the need for credentialing trademarks. My thinking is that U.S. Department of Transportation or a transportation industry association could acquire the trademarks from me once they see the importance of credentialing the infrastructure and vehicles.

For the time being, I think these terms are in better hands with me. That's because I am committed to Vision Zero, carbon emissions reduction, public health, social equity, sustainability, economic opportunities, free access, and more. These principles are baked into my trademark plans. Essentially, these are defensive trademarks to protect these terms from bastardization. At some point, I hope to transfer them to a larger entity with the condition that these credentialing principles are upheld.

Tiny Transit, Megatrend

The ideas in this book are neither liberal nor conservative. They are neither Democratic nor Republican. This movement is about building common ground to forge a broad movement that is united and unstoppable.

"How lucky we are to be alive right now," is the line from *Hamilton*. We are truly lucky, all of us. We are plugged into one of the biggest megatrends we will see for some time. I hope that you appreciate this extraordinary moment in American history and that your mind is whirling with ideas and possibilities.

The only things standing in your way are inertia, inaction, lack of resolve, silo thinking, blaming the process, and the deadly "that's not how we do things."

I recall my conversation with the professor who introduced me to Katie, effectively launching this project.

"Won't automakers try to kill this?" I asked.

"No, not anymore," he said. "They know something like this is coming. They just want to know what it is."

"Won't big oil kill this?" I asked.

"No. *Young people won't let them this time.*"

It is up to cities to lead the way. Let's combine efforts to create LEAN networks and unleash their power to cut carbon emissions.

Scooter enthusiast Sean Buggs enjoying his day

Chapter 12:

Putting it into Action

Henry Ford said, "If I'd asked customers what they wanted, they would have told me, 'A faster horse!'"

People had to see and experience the first automobiles before they wanted to own one. Once automobiles were out there, it was like the physics demonstration where you toss a ping pong ball into an arrangement of balls, and over the next few minutes that first toss sets off more ping pong balls until you have an unstoppable chain reaction.

In the same way, people first need to experience NEVs.

Introduce NEVs with Hands-On Test Drives

No matter how hard you talk or try to explain NEVs, words and pictures are inadequate. The most compelling case will always be to place someone in the driver's seat and offer them the chance to test drive an NEV. From then on, it is a chain reaction as people become converts and evangelists for NEVs.

It's not hard to do because there are people who will help you do it. We've successfully staged demos with support from Polaris GEM and their talented team. You can see people transform before your eyes. Before, most have no idea what an NEVs is. Chances are very few people in your cities will have heard of NEVs. Even people who claim to know are often thinking of golf carts. For nearly everyone, this will be their first introduction to NEVs.

Think about places people gather: farmer's markets, festivals, shopping malls, city halls, sports events, neighborhood meetings, places of worship, arts and cultural events, and conferences. Then contact member cities that are open to innovation. They can help and may cohost it with you. You don't have to stage an event, publicize it, and get people to come. The event is already staged, and people will already be there.

This is the time to tap the "public information officers" (PIOs) on staff with your COG, MPO, or member cities. PIOs are great at creating demos attached to other events. PIOs can identify the best and most feasible opportunities, handle publicity and social media, take photos and video, get permissions to feature people in photos, and work out the arrangements and logistics.

Basically, go where the people are. Enlist PIOs and member cities. It's easy, fun, and risk-free. You know people will show up because you're meeting them where they will be.

Remember, you can't just drive an NEV to these events. In most states, you cannot drive an NEV on a road where the posted speed limit is forty-five or higher – and you shouldn't even do that. Drivers aren't watching for these vehicles. The roads are already dangerous enough without adding unfamiliar slow-moving vehicles to the mix.

But you can turn this problem to your advantage. Ask about getting a police escort. See if your PIO to help get social media on it. You can use the opportunity to introduce LEAN networks – the reasons they're needed and the problems they solve. It's a way to inform and educate people.

You can multiply the education quotient by including other low speed modes in future demos. Introduce pedicabs, scooters, seated scooters, bicycles, pedal assist bicycles, Scoots, Segways, even grown-up trikes – all modes that are growing in popularity but that need a protected network to ensure their safety.

Hands-on events provide terrific opportunities to get to know the market for this new concept. Conduct research. Talk with people. Understand where their thinking is and what appeals to them about NEVs.

Polaris GEM has provided an email address especially for readers of this book to contact the company directly to arrange a demo and learn more about more about GEMs. Contact: GEMsustainandsave@polaris.com and mention Tiny Transit.

Tricycles for adults can carry passengers

Tooling around on a beautiful day

Choosing Demonstration Projects

In Austin, South Lamar Boulevard is a busy street. People have suggested that it would be a good location to pilot LEAN networks. Certainly, it needs a solution for congestion, but it's probably not a good candidate for an introductory demonstration project. The road is four lanes with a center turn lane and constant traffic. There are curb cuts all along the road, many close together. Visibility is poor, addresses are poorly marked, frustrated drivers make risky left turns and U-turns, and the road has many distractions. Drivers make quick decisions to swerve around the frequent buses. There is no shoulder. Dodging a car could mean swerving into a bicyclist. The buildings are close to the road. Small businesses struggle anytime there is road work. It would be a locational challenge.

The people who live and work along South Lamar Boulevard desperately need a mobility alternative, but there are so many compounded challenges that I don't see how it could be a demonstration site for LEAN Lanes.

It's better to start with locations with fewer challenges.

In Chapter 6, I described how Katie and I worked over the course of three days to identify possible demonstration locations in Austin. We wanted to show specifically how LEAN networks could serve different needs in different parts of the city. One location was a small, private university campus that could evolve to become a car-free zone and extend across a busy street to student housing. Another would connect two neighborhoods with transit and grocery shopping. A third would connect a neighborhood with a community college campus that's expanding. We then prepared maps and a photo album.

I want to encourage you and your cities to approach this activity in somewhat the same way. You don't need a six-month study. This process is to *accelerate* adoption, not slow it down. Your purpose is to engage people in bringing their thinking to the process. It is to develop and float ideas and refine them, not bat them down. The objective is for communities to identify possible demonstration project routes that will appeal to others.

Cities should aim to identify several different types of LEAN applications to help people envision the variations possible in a new approach to mobility. Here are ideas you can use to select candidates.

1. Is there an overly wide road with excess capacity sufficient for LEAN Lanes?

2. Is there a new transportation corridor being planned into which LEAN network could be incorporated?

3. Is there a college or university campus where low speed modes could help transition the campus into a "no car zone?"

4. Is there a transit center where a first/last mile solution could improve access and increase ridership?
5. Is there a neighborhood where streets have excellent visibility and lower speeds on residential streets would be acceptable?
6. Is there a "new urbanism" development on the drawing board?
7. Are there off-road paths that could be expanded and extended to link destinations?
8. Is there a wide median along a highway that could accommodate LEAN Lanes?
9. Is there a development where the developer would be willing to work with transportation engineers to include LEAN networks in exchange for other concessions?
10. Is there a parking structure at the design phase that could be flexibly designed to serve low speed modes as the transition to LEAN networks occurs?

Remember, your work product is a communications tool to stimulate thinking, not a formal plan. It begins as a work in progress. It should lead to a demonstration that is such a success that it calls out to be replicated.

Strategic Mobility Plans

Ideally, your organization and your member cities will incorporate LEAN networks into their strategic mobility plans as a way to reduce carbon emissions and get out of, or stay out of, nonattainment. These plans set future directions.

Here are actions your cities can take to incorporate this new mobility alternative in their strategic mobility plans.

- **True Vision Zero.** Make it clear in your cities' plans that Vision Zero is not an aspirational wish, but an absolute requirement for your LEAN network. Invite private sector safety experts and smart city technology experts into your process. Adopt safety measures coming out of the gate. Networks for low speed, low emission vehicles must be protected. Painted lanes on the edges of arterial roads are insufficient, just illusionary. The plan cannot allow NEVs and their drivers and passengers to come into conflict with a two-ton vehicle whose driver drifted into a LEAN Lane while glancing at his iPhone. People could be killed. And if you don't win this point at the outset, all of the potential carbon reductions that could come from the LEAN infrastructure will be jeopardized.

- **Recognize NEVs.** If NEVs aren't recognized as a mode in your COG or MPS's mobility plans, work to add them as a start. Hands-on test drive demos with your member cities will help get them on board.

- **Recognize all low speed, low emission modes.** People are transitioning to new, low cost independent mobility modes. In Dallas, Texas for example, there were 750,000 trips on dockless scooters in just the first few months after they came to town. This rapid adoption changes everything. Have you heard the phrase "people voting with their feet?" We now have people who are voting with their wheels. Easy, low cost modes aren't going away.

- **Recognize your city's authority.** Cities can limit the color palette for NEVs and all low speed modes and require that low speed modes are in neon colors that are super-bright and reflect light. Cities can require that they display the low speed triangle like many low speed utility vehicles do. Take a page from Uber. Its red neon dockless bikes make them noticeable even on streets where there is a lot of competition for drivers' attention. We wouldn't think of letting a highway worker go without a neon vest. Why would we treat young people with any less concern for safety than the highway worker?

- **Establish safety practices without delay.** We are now beginning to see injuries from scooter falls and collisions. Lawyers are advertising on billboards: "Injured on a Rented Scooter?" Take the guesswork out of what is allowed. We're now seeing an epidemic of "jay-scootering" and near misses by cars. To allow it is to sanction it. Do we really want to do that? That is where inaction leads.

- **Ensure that data is collected on every incident** of any injury or crash. You may think that cities collect this information, but I have yet to meet one that does. In the case of the cycling hit and run, the police couldn't open a file until the bicyclist filed a police report. He didn't. That tribal knowledge was lost. We can and must provide apps so that people can report them, and the app captures the location. Without this information, we are simply biding our time until a tragic accident occurs.

- **Look to innovator and early adopter cities for ideas from their strategic mobility plans.** You will become more knowledgeable fast by studying their websites for how they have incorporated NEVs and low emission modes in their plans. LA Metro's "Slow Speed Network Strategic Plan for The South Bay" has useful visuals. Peachtree City's plan makes off-road paths a cornerstone. In each plan, you might find something relevant to your region. You'll find links to these and more on our Tiny Transit Strategies site, TinyTransit.com.

Strategic mobility plans are simply words and images on a page or screen. But these plans are terribly important. The cities you work with will probably not develop LEAN networks as a standard practice or even as a pilot project unless they appear in their strategic mobility plans.

LA Metro's strategic network plan

Starting with Community

One of the wonderful things about Tiny Transit is how you can see people change before your eyes as they experience Neighborhood Electric Vehicles (NEVs). It is a transformational experience. I still recall my own wonder when I first rode in one, an eCab in downtown Austin. The lightness, the stripping away of the tons of metal to the essence of the ride. People aren't hidden behind tinted glass or sitting six feet off the ground in an SUV. NEV drivers are at the same level as pedestrians.

Steve Jobs once said, "If you want to make people happy, sell ice cream."

I know just what he meant. One of my summer jobs was behind the counter at a popular ice cream chain. I seldom saw a dour person. And whatever rainclouds might have been dragging people down evaporated as people began tasting flavors they'd never imagined before. German Chocolate Cake ice cream…?! Lemon Peel ice cream…?! They were all smiles. I felt just as thrilled for them. They were so *happy*.

Ice cream is a social equalizer. No amount of money can buy a better ice cream than the other customers have. For very little money, a person can buy a scoop of the very same ice cream that billionaires buy. The only stresses are choosing a flavor and wondering if there is a limit on how may tastes you are allowed.

There are no knife fights at ice cream parlors. No one tries to shove ahead of another customer. No one is making obscene gestures. People are patient and polite. They take their time making selections, sharing tastes. Messy kids are adorable. No one is worried about stains or clean up. There are bright colors, bright lights, bright moods.

NEVs are the transportation equivalent to selling ice cream. People relax, their faces light up, and the world feels new and fresh.

It is hard to image road rage among NEVs drivers. You see drivers as *people*, not vehicles. In NEVs, you can actually recognize faces, make eye contact, and indicate your intent with a friendly wave or hint of a smile. Without the anonymity of tinted windows, drivers are politer.

NEVs foster a code of behavior more like that at a neighborhood gym than at Black Friday morning at Walmart. They may very well be the same people at both, but NEVs bring out our better selves.

All of this people seem to instinctively understand once they have a chance to experience NEVs.

LEAN networks strengthen community. Communities must be a part of planning.

The Better Block is a nonprofit based in Dallas that shows community members that they have the power to make changes in their neighborhoods, and they show City Hall how these changes would work. "Better Blocks" are speed dating for communities, and they're just the first step. Once you've had a better block experience, then you have identified the leaders in the community and they've been organized." Ideally, pedestrians are safer, businesses are revived, and neighbors know one another.

Better Block projects are a model for cities to work with neighborhoods to develop LEAN demonstration projects.

Chapter 13:

Overcoming Obstacles

I recently told a friend about this book. He asked, "Won't people then take your ideas? You've just told them how to create LEAN networks. They won't need you."

I only wish it were that easy.

Anytime you're introducing a paradigm change, there will be challenges. Changing how we think is the greatest obstacle of all. Even though we are talking about a new and parallel physical infrastructure with LEAN networks, the greatest challenge is not physical.

"I think the greatest challenge is 'that's how we've always done it,'" said Michael Rogers, Director of Transportation for the City of Dallas. "The challenge is the mentality, 'This roadway was designed for my vehicle, not for walking, not for the disabled, not for those riding in transit, not for bicycles, not for other types of vehicles.' I do not think that this challenge is insurmountable. It's a challenge that we need to take up."

This challenge will stall the transition to low emission, low speed, low cost modes until people realize that the way we've always done it is no longer in their best interests.

We need to listen carefully to what is behind people's positions. We have to engage with people who are firmly entrenched and respect their reasons for questioning or opposing change. We have to recognize that on many points, they're right and acknowledge that. We have to spend authentic time with them. Then if we have to prove them wrong, we do it in a magnanimous and respectful way, a way that recognizes their everyday heroism, and welcome them into the big tent.

It sounds like it takes a lot of time, right? In a sense, yes.

But how much time does it take for city officials to sit through a public hearing where opponents turn out to filibuster at the microphone?

How much time does it taken for a community to delay a pilot project because the support is not firmed up?

How much time does it take for acquiring rights-of-way when it may be possible to have neighbors willing to contribute it because LEAN Lanes will serve and benefit them?

City leaders have to spend the time, and you are in a role to help them. As you will become increasingly knowledgeable, you'll see the connections among your cities, ways they can share best practices and site-specific solutions.

Obstacles to Anticipate

Let me run down a list of obstacles your cities might encounter and how they prepare to surmount them.

1. *"Providing for NEVs will never make a dent in congestion. Most people will never use these modes."* Nothing you can say

will convince someone differently. Even the explosion in dockless scooters and bicycles may not convince someone who is adamantly opposed to change that these new modes are more than a "pet rock." The way to overcome this mindset is with successful demonstration projects.

2. *"NEVs and scooters cannot be made safe."* Elicit their specific concerns – maybe they knew someone who was killed while riding a bicycle. LEAN networks are the answer. We recognize that these vulnerable modes must be protected in order to truly take off.

3. *"Only the affluent will be able to afford NEVs."* Ask, what is their rationale? Perhaps they see only the sticker price of $10,000+ for new NEVs and don't realize that the five years' cost is about one-quarter of the cost of conventional vehicles. Show that you can buy a good quality used NEV for around $4,000 to $6,500. And there are possible collaborations to provide financing for people with low credit ratings. Local banks and financial institutions could make good partners on this, pooling funds to make these loans. In Austin, some tech employers are now offering mobility packages that an organization called Movability helps to develop. And of course, other low emission modes cost a fraction of what NEVs cost.

4. *"This change will hurt small businesses that serve the auto industry."* There is some truth to this for auto collision repair, towing companies that get cars off the road when they're totaled, gas stations, hospital trauma centers, and mortuaries. But this is a problem we hope to have. Just as cities offer gigantic "packages" of incentives to woo major corporate relocations with their thousands of jobs, it might be wise to provide practical support to small businesses

that want to transition to the new mobility over the next several years.

5. *Despite good intentions, there will be runaway costs, and government will screw it up.*" This is the "it sounds too good and I don't believe it" argument. It's the $12,000 toilet seat. There are days I fall into this camp. But here is what is possible as we design this new mode. We can use big data to inform us as we proceed. We can use tribal knowledge to build the $60,000 small bridge. We can incorporate cost-saving smart technologies. We can track costs with the intensity we track our own personal purchases. We can spend public money just as we spend our own – very carefully. If we cannot demonstrate the competence to watch our spending, why should governments be trusted to build larger projects? I recommend tapping this discontent to fortify initiatives so that they deserve the public's trust.

6. "*This is for millennials, while the rest of us have to work with our hands. We need F150s. And now you're taking away our lanes.*" I understand this, too. I call it "mode rage." This is why I don't like preferred parking as an incentive for low emission modes; it's irritating. How annoying it must be to see affluent young people zipping past on their scooters, to see preferred parking for electric cars as if they are disabled, wondering if city leaders have lost touch. But this is surmountable, too. We need to hear these complaints and resentments and respond in how we implement LEAN networks. When we make safety the prime directive, when we measure dockless trips in terms of the congestion they eliminate, when we have earned the trust of conventional vehicle owners, we will earn public trust.

7. *"That darn EPA."* The only way I know to combat this is to develop LEAN networks that prove that they can be built with the crumbs of major transportation projects *and* reduce congestion. Use nonattainment, or fear of nonattainment, as platform for improving lives in all pockets of our cities. EPA or no EPA, LEAN networks support economic development and resilience.

8. *"This is a liberal idea."* Not at all. This objection is why it is so important to have respected conservatives in your big tent. It's why I have a mixture of political and socioeconomic perspectives as my sounding boards. We don't have to agree on everything. But we need to be shoulder-to-shoulder on this. LEAN networks are fiscally conservative and market-driven. They allow low- and moderate-income people to lift themselves up and improve their lives. LEAN networks go straight to the heart of two of the Pentagon's assessments of the greatest threats to our national security: climate change and our dependence on foreign oil.

9. *"This will raise taxes. Our city can't afford it."* To the contrary, your cities can't afford *not* to do this. The LEAN approach, by relieving congestion with low cost solutions, will eliminate the need for more taxes to build more gargantuan road and highway projects. It will make public transit projects more responsive to demand, saving money as well. Speaking as a Texan, property taxes are killing us. This is the first hope on the horizon for containing them.

10. *"Get in line behind everyone else. You want something, you've got to wait your turn."* You can't win over this thinking; I've tried. You have to go around it. Know this will happen, and don't let it discourage you. Harness your reaction to move forward.

11. *"A protected network would involve considerable infrastructure, investment, public engagement, consideration of use of right of way and economic impacts, and policy change."* This is the big "no." The thinking is that it is just too much and will never happen. This might come from people who don't want to enter what they fear is uncharted territory. The only way to turn this around is for COGs and MPOs to be a proactive resource for information that will help bring innovation-minded cities up to speed on what other cities are doing across the country.

Building consent requires patience. I understand that people can feel threatened by paradigm change, so I tread with care. We need to tear down silos that are obstructing change, but we can be sensitive and respectful about it.

Strategies for Overcoming Obstacles

Most people are quickly intrigued by the LEAN concept, want to know more, and are willing to help in whatever way they can. By and large, the LEAN concept wins over ninety-plus percent of the people we meet with. Moreover, we get valuable feedback from one hundred percent.

Once in a while, someone will try to discourage us. One man poo-pooed our ideas. I said gently, "I'm surprised because most people tell us that they like what we're doing." He responded, "I think those people were just being nice to you two ladies." That was passive-aggressive. I recognize that he just doesn't understand – yet.

Laughing turns a hurdle into a speed bump. Honestly, did someone really think that saying, "You ladies don't have a scalable business model," is going to make us give up? Who ever said we wanted that? And why is it that the few who try to discourage us call us "you ladies?"

Let me share some strategies that have worked for me with those few people who try to discourage us.

When people tell us "no" – usually in the context of government pushing out decisions that could change the status quo and thus affect their work environment or people who don't like that we're tearing down their carefully fortified silo walls – we don't give up. We don't even give up on *them*. I assume everyone can be won over; they just don't yet realize we're not going away. We have no enemies, nor should you. We only have people who don't know us well enough to like us – yet.

I'm reminded of a movie scene where a gangster is shocked to discover someone he thought he had murdered now in his office, leaning back in his very own chair, saying, "Remember me?" I've seen that look on the gangster's face on people who assumed that after their crushing assessment, we'd give up – but here we are, spirited as ever. Often our idea is better than before. I have good recall on each conversation, and when possible, I show them how we've used their feedback to improve our concept.

When people tell us "no," but I know the answer should be "yes," I realize that I just haven't communicated well. Or perhaps we've just met them when they were having a bad day. I don't let things bother me. I don't have the time for emotions like irritation, anger, or dismay. I have work to do.

I know that the way to win difficult people over is not with verbal jousting but with proof of concept – in this case, successful demonstration projects. I assume that down the line, they will come to love LEAN networks so much that the memory of their negativity is erased, and they will recast themselves as early advisors.

I have seen this phenomenon before. It's the natural evolution of the big inclusive tent.

Never forget how fortunate you are. Your work is meaningful. It can improve lives and health and maybe save our planet. Improving air quality matters. Never forget that you and the cities you work with have the power to change the course of our world.

Steal This Book, Please!

You will face obstacles and discouragement because it's human nature to oppose what we do not understand. Most of these obstacles are based on fear or misunderstanding.

Your big tent is the key to overcoming objections. You may not win everyone's active support, and that's okay. You might rather have a beautiful unanimous consensus – who wouldn't? But that's probably not possible, and I'm not sure I'd believe it. Some people are naturally critical and cantankerous, but they're still part of your big tent. They may have important things to say, for you to hear. Grudging consent is a win.

I invite you to take these ideas and make them your own. In the words of Abbie Hoffman, "Steal this book – please!"

Chapter 14:

Quick Start Guide

How will you and your cities know when they are on course? The following metrics are a sample checklist that cities can adapt for their unique circumstances.

1. Your city will have a **strategic mobility plan** that recognizes the need for citywide plans to safely accommodating NEVs, pedicabs, dockless, and other low speed modes.

2. Your city will have its own **"crumbs report"** that identifies examples of how LEAN networks can be built with the crumbs of major transportation projects.

3. Your city will offer **hands-on demos and test-drive events** where people gather – to build awareness, educate the community, and lay the foundation for a groundswell.

4. Your city will have a **big tent** where people with diverse reasons for wanting this mobility alternative are welcomed.

5. Your city will have identified **possible demonstration projects**, with at least one underway.

6. Your city will have a **real estate developer** proposing to include this mobility alternative in their projects, and **city planners** ready to work with such a developer.
7. Your city will have a **right-of-way donation** plan that encourages property owners to expedite service to their property with protected and off-road paths. Donated ROW fosters the right spirit for this mobility alternative.

Quick Start Game Plan for Cities

You can help your cities approach this low emission mobility alternative step-by-step with this quick start game plan. Your help will be especially important to cities that aren't large enough to have what they need in-house.

1. **Host hands-on public demos**. This comes first. A test drive is worth ten thousand words. There is no experience quite like getting behind the wheel of an NEV. It's transformative. People spontaneously want to welcome this new mode. Remember to involve your public information officers. You can start now by arranging a demo with Polaris GEM. Email GEMsustainandsave@polaris.com.
2. **Identify cities that have the potential to become innovators and early adopters**. You may already have a sense of which cities and communities may be most receptive. Some people will stand out. They start conversations and are eager for more. Your job is to identify and discuss LEAN networks with these prospective early adopters.
3. **Start your big tent with public health and public safety leaders**. These will be some of your most passionate and effective allies. These include police chiefs and public health officials. Involving these leaders will help ensure practices

that prioritize safety such as LEAN protected network design, low speed vehicle signage, and neon colors.

4. **Inventory new roadway construction** still in the design phase. While there's still time, your cities can consider pending projects for LEAN networks. Is a city is spending millions of dollars on roadway expansions or smart corridor projects, for example? If so, including LEAN networks and connections just makes sense.

5. **Inventory planned developments**. Likewise, what proposed developments are still on the drawing board? Would the developer be interested in incorporating the LEAN infrastructure if it did not add to the cost or schedule? Better yet, could incorporating LEAN infrastructure *reduce* costs and make the development more profitable?

6. **Initiate "crumbs reports."** Cities can use the report I wrote for Austin as a prompt for their own versions of "How LEAN Networks Can Be Built with the Crumbs of Major Transportation Projects." These reports should change the conversation from "We can't afford this," to "We can't afford *not* to do this."

7. **Include NEVs and protected networks in strategic mobility plans.** Are NEVs and other new low speed modes included? Ask your cities, what do they need to bring their strategic mobility plans up to LEAN thinking? It may start with one small step; for example, resolutions by their governing bodies to recognize NEVs.

8. **Help cities identify possible demonstration projects.** They can start communicating with a map and photo album to help people see possibilities where LEAN networks could be implemented first. This will generate conversations,

and everyone will have their own thoughts to share. From this activity will come the initial demonstration project candidates. At TinyTransit.com you can find direction on identifying successful demonstration projects. You and your member cities can become part of the online community of innovation-minded cities by contacting me at the Institute for Community MicroMobility.

9. **Celebrate**. You and your cities will meet milestones along the way. Celebrations are bonding. Being an innovator city fosters a sense of pride. Everyone in the community can be a part of the solution to carbon emissions.

Paradigm changes aren't easy. They can take years. We don't have that kind of time. Cities can't wait on Washington D.C.

Fortunately, most cities are nimble, motivated, grassroots-minded, and under the gun on nonattainment. They are the point source for change, at the heart of an emerging groundswell.

"The neighborhood is the unit of change," says David Brooks, conservative columnist for the *New York Times*. "If you're trying to improve lives, maybe you have to think about changing many elements of a single neighborhood, in a systematic way, at a steady pace." He could very well be talking about LEAN network demonstration projects.

Once innovative cities understand what is possible – a low speed, low cost, low emission mobility alternative with many additional benefits – these cities will join in accelerating this mobility alternative.

Let's begin. Let me hear from you. This paradigm change can begin now, with you.

Helpful Resources

- **Institute for Community MicroMobility** is a nonprofit that I founded as an online community resource.
- **Tiny Transit Strategies** offers information, research and links to other resources.
- **Polaris GEM.** To arrange a demo, contact: GEMsustainandsave@polaris.com.
- **Los Angeles Plan**. See LA Metro's *Slow Speed Network Strategic Plan for The South Bay*.
- **US Ignite**. See its Smart Gigabyte Communities program.
- **Yale Program on Climate Change Communication** offers maps and studies of public opinion on climate change.
- **Audubon**'s Birds and Climate Change Report is a call to action.

Acknowledgements

I'm filled with gratitude for Dr. Pamela Ryan and the Tingari-Silverton Foundation for their belief in the impact we could have. And there would be no book without Katie Kam, Nan McRaven, Gail Papermaster, Valorie Keller, Jennifer Failla, Catherine Crago Blanton, Marie Crane, Dane Anderson, Sue Kolbly, Michael Walton, and Howard Falkenberg,

Thanks to my "sounding boards" who kept me moving forward: Phoebe Allen, Sarah Campbell, Bre Clark, Gerald Daugherty, Priscilla Douthit, Gary Farmer, Joene Grissom, Emily Little, Tobin Levy, Joan Marshall, Congressman Michael McCaul, De Peart and the Downtown Austin Alliance, Lisa Kay Pfannenstiel and Movability Austin, T. Paul Robbins and the Austin Environmental Directory, Michael Wheeler, Laura Bond Williams, John Wooley, Don Young, Ida Zamora, and Stacy Zoern Goad. I especially appreciate my friend since third grade Linda Wurzbach who provided an office where I could work uninterrupted.

I am grateful to the people in the micromobility industry who shared their knowledge with me: Chris Nielsen, Electric Cab North America; David Knipp, Movemint Bike Cab; Russ Ziegler, Sean Cheatham, and their team at Polaris GEM. Just imagine how patient they must have been with me. I was so new to it that when I started, I thought NEV stood for Nevada.

Thanks to Rob Spillar, Jason JonMichael, Karla Taylor, Joel Meyer and the City of Austin transportation team for spending time with me and for their willingness to help educate me.

Thanks to the City of Dallas: Michael Rogers, Director of Transportation, Marva Fuller-Slider, Chelsea St. Louis, Maria Chadwell, and Brita Andercheck.

Thanks to inspiring California innovators Tom Kirk, Coachella Valley Association of Governments (CVAG); Marissa Mangan, San Diego Association of Governments (SANDAG); and Ray Leftwich, City of Lincoln.

I am immensely grateful to all of the people who were willing to meet and share their thoughts over the past three years: Austin Mayor Steve Adler, Adrian Shelley, Angelos and Esther Angelou, Amy Atchley, Lucia Athens, Thais Austin, Kris Bailey, Chad Ballentine, Dianne Bangle and the Real Estate Council of Austin, Zach Baumer, Magnolia Blue, Anne Boysen, Carolyn and Cap Brooks, Juanita Budd, Sean Buggs, Casey Burack, Brandi Clark Burton, Jim Butler, Bruce Byron, Kevin Chandra, Randy Clark, Lee Cooke, Anton Cox, Dan Dawson, George Cofer, Lee Cooke, John-Michael Cortez, Ken Craig, Cassandra DeLeon, Gordon Derr, Marion Douglas, Sue Edwards, Peter Einhorn, Louise Epstein, Kyran Fitzgerald, Paulette Gibbons and ULI's Austin chapter, Cristal Glangchai, Bobby Godsey, Alan Graham, Tricia Graham, Sam Graham, Sherri Greenberg, Sandy Guzman, Kris Hafezizadeh, Jeff Hahn, Hailey Hale, Stephanie Hayden, Clarke Heidrich, Todd Hemingson, Louis Henna, Andrew Hoekzema, Alan Holt, Bob Honts, Ora Houston, Michelle Huntting, Joe Ianello, Celia Israel, Parker Jarmon, Barbara Johnson, Katherine (Kat) Jones, Dick Kallerman, Carly Keller, City Council Member Ann Kitchen, Laraine Lasdon, Mike Levy, Austan and Diane Librach, Emma Lou Linn, Lizette Melendez, Lynn and Tom

Meredith, Nick Maynard and US Ignite, Bob Metcalfe, Art Mitchell, Katharine Mitchell, Nick Moulinet, Bob Moore, Peter Mullan, Luke Mulvaney, Alan Nalle, Chris Nielsen, Mike Nellis, Krista Nightengale, Jackie Nirenberg, Marc Ott, Karl Popham, John Rigdon, Mike Rollins, Alix Scarborough, Michael Searle, James Sharp, Brigid Shea, Ted Siff, Beverly Silas, Janel Simms, David C. Smith, Ingrid Spencer, Robin Stallings, Dan Sturges, Geoffrey Tahuahua, Jeff Thomashow and Scoot, Donna Tiemann, Jeff Travillion, Roy Truitt, Michele Van Hyfte, Lesley Varghese, Jeff Vice, Kirk Watson, Tod Wickersham Jr. and Beneficial Results, Phil Wilson, Anne Wynne, and Peck Young. There may be people among these who've even forgotten our meeting and can't remember what they said. But I remembered, and it mattered.

Thanks to the gifted photographer and designer Roj Rodriguez.

My deep appreciation to the founder of The Author Incubator, Angela Lauria, and her team, including Bethany Davis, who coached me to write a book that will make a difference.

I also wish to acknowledge the terrific Morgan James Publishing Team: David Hancock, CEO & Founder; my Author Relations Manager, Gayle West and special thanks to Jim Howard, Bethany Marshall, and Nickcole Watkins.

Finally, I am most grateful to Tyler and Jack Rosenberg for their love, imaginative ideas, and willingness to live on fumes while I have pursued my passion for Tiny Transit. My hope is that readers of all ages will be inspired and empowered to join in the groundswell for change. Greta Thunberg, thank God for you and the millions of children you inspire. Let's tear down those silo walls and embrace this paradigm change together.

I would love to hear from you. Find me at tinytransit@gmail.com.

Thank You

I am delighted that this book made its way to you. In appreciation, I am making available to you a webinar, "LEAN Networks: Improve Mobility While Cutting Carbon Emissions." This is especially for mayors, activists, city leaders, Councils of Government, and Metropolitan Planning Agencies who are helping cities reduce carbon emissions with the right sense of urgency.

Please email me at tinytransit@gmail.com, and I will gladly provide the link for the webinar. You can also sign up for email bulletins on what is happening in cities and regions across the country to provide protected networks for NEVs and other low emission modes.

I offer a thirty-minute free phone consult specifically to welcome your questions and thoughts. I'd love to hear from you.

If you know a classroom, neighborhood organization, or nonprofit group that could benefit from this book, please let me know so that I can make the book available to them at little or no cost.

About the Author

Susan is the founder and executive director of the Institute for Community MicroMobility, a nonprofit resource for innovation-minded communities.

Susan founded Tiny Transit™ Strategies, an Austin-based grassroots movement that advocates, educates, and consults on protected networks for Neighborhood Electric Vehicles (NEVs) and other low speed modes as a low-cost urban mobility solution. This Low Emission Alternative Network (LEAN) infrastructure has the potential to relieve traffic congestion, reduce traffic fatalities and life-altering injuries, improve the social determinants of health, reduce financial stress, reduce air emissions, reduce the impact of transportation on the environment, help cities grow their economies, reduce property taxes, and improve mobility for everyone. For cities, this concept is a game changer. For the nation, this new transportation alternative is a step toward economic resilience, reduced carbon emissions, and energy independence. The LEAN concept was one of seventy out of three hundred applications chosen for a national Smart Cities Connect Innovation Showcase in 2017.

Susan's background is in economic development. She served as project manager or senior editor for each of the three long-range economic development plans for Austin, Texas spanning twenty years, 1990s–2000s. She was aide to Austin City Council Member Lee Cooke, who later became mayor. At the City Council, then as a director in the Economic Development Division of the Austin Chamber of Commerce, Susan was one of the leaders involved in envisioning and positioning Austin as having a key role in the global information economy. She led the Chamber's National Marketing, Target Marketing, and National Public Relations programs through which Austin broke into "Top 10 U.S. Cities" cover stories in *Fortune, Forbes, Parenting,* and other periodicals, leading to a cover story in *National Geographic.* Her articles on the role of universities in economic development were published in national journals.

Susan is founder and president of Engelking Communications LLC, a marketing and public relations firm that since 1998 has consulted with dozens of clients in an array of industries, including technology, education, health care, utilities, real estate, and nonprofits. She produced an annual report that was named one of the top ten annual reports in the U.S. by the Stevie® Awards. She serves as strategist, ghost writer, speechwriter, editor, investor communications counsel, and trusted advisor for her clients. With journalist Dane Anderson, Susan cofounded WordMuscle, a writing and communications strategy firm.

She is a founding shareholder in EE-ABF Holding LLC, Texas Pyrolysis Group, and Earth Energy Renewables, all in the sustainable waste recycling space and based in Bryan, Texas.

Susan served three terms as president of the Austin Children's Museum, raised over $3 million for its expansion, and was instrumental in its evolution to become the Thinkery. Susan has been named Austin Communicator of the Year by Women Communicators of Austin. Susan was treasurer for Nan McRaven's winning campaigns for trustee for Austin Community College.

Susan has a master's degree from the LBJ School of Public Affairs at the University of Texas at Austin and B.S. in education, also from UT Austin, *magna cum laude*, with concentrations in English and history.

Susan has children in high school. Her prime directive for Tiny Transit™ Strategies is safe, low speed, low cost, low stress, low emission, climate-conscious mobility for their generation. She doesn't want to see another young person killed on our roads.